# AIDS

# AIDS
## Taking a Long-Term View

**The aids2031 Consortium**

Vice President, Publisher: Tim Moore
Associate Publisher and Director of Marketing: Amy Neidlinger
Acquisitions Editor: Kirk Jensen
Editorial Assistant: Pamela Boland
Development Editor: Russ Hall
Operations Manager: Gina Kanouse
Senior Marketing Manager: Julie Phifer
Publicity Manager: Laura Czaja
Assistant Marketing Manager: Megan Colvin
Cover Designer: Gary Adair
Managing Editor: Kristy Hart
Project Editor: Betsy Harris
Copy Editor: Krista Hansing Editorial Services, Inc.
Proofreader: Kathy Ruiz
Senior Indexer: Cheryl Lenser
Compositor: TnT Design
Manufacturing Buyer: Dan Uhrig
© 2011 by Pearson Education, Inc.
Publishing as FT Press Science
Upper Saddle River, New Jersey 07458

ISBN-10: 0-13-217259-3
ISBN-13: 978-0-13-217259-2
Pearson Education LTD.
Pearson Education Australia PTY, Limited.
Pearson Education Singapore, Pte. Ltd.
Pearson Education Asia, Ltd.
Pearson Education Canada, Ltd.
Pearson Educación de Mexico, S.A. de C.V.
Pearson Education—Japan
Pearson Education Malaysia, Pte. Ltd.

Library of Congress Cataloging-in-Publication Data
AIDS : taking a long-term view / the aids2031 Consortium.
      p. ; cm.
   Includes bibliographical references.
   ISBN-13: 978-0-13-217259-2 (hardback : alk. paper)
   ISBN-10: 0-13-217259-3 (hardback : alk. paper)
   1. AIDS (Disease)  I. aids2031 (Organization)
   [DNLM: 1. HIV Infections—prevention & control. 2. Acquired Immunodeficiency
Syndrome—prevention & control. 3. Health Policy. 4. International Cooperation. WC 503.6]
   RA643.8.A435 2011
   362.196'9792—dc22
                              2010031031

# Contents

# Preface

There are a number of compelling messages in this important book, *AIDS: Taking a Long-Term View*. The over-arching message is that the world needs to come to terms with the fact that AIDS is not over, by any measure. Over 25 million people have died from AIDS since it was first reported in 1981, and more than 33.3 million are living with HIV today. We have made remarkable progress against the disease, with greatly expanded access to prevention and treatment in just the last few years. AIDS is, nevertheless, an enduring, generations-long challenge that can only be effectively addressed with long-term thinking, planning, and investing.

Our understanding of AIDS has matured over the course of the past three decades since it was first reported. Many no longer see AIDS as a global emergency threatening everyone in every country and at every level of society. While in some countries AIDS has become hyperendemic, such as in Southern Africa, in others its most devastating effects have become concentrated and entrenched in the most marginalized communities.

To this far more heterogeneous epidemic landscape, aids2031 brings fresh thinking. We need to look up, look farther ahead, and think seriously about the long-term implications of our actions today. For the first three decades of AIDS, our response has been exactly that—a response—a reaction to a global emergency that needed immediate attention with little information about the disease and its origins. Today, we have far more information at hand, but are we using it? We cannot give up the quest for new knowledge—new prevention tools, better and more efficient treatment, better understanding of the social drivers of HIV transmission, and better evidence of what works and what doesn't and why. But, while we work toward improving our knowledge, we need to make much better use of the information and tools we have in front of us. Without more efficient prevention efforts, new HIV infections will continue to grow; without more effective treatment strategies, people will continue to die.

As leaders in the continuing global fight against AIDS, we recognize that there are a number of challenges ahead—some persisting, some new, and some which are yet to surprise us. We welcome this important analysis and recommendations from aids2031, an independent initiative to take stock of where we are today, and consider what we need to do better, or differently, to dramatically reduce the number of HIV infections and AIDS deaths by 2031, 50 years after AIDS was first reported. We laud the process of aids2031 which, through its nine thematic working groups and its young leaders network, has involved more than 500 people around the world. Indeed, the process has been as important as the products.

We encourage you to read *AIDS: Taking a Long-Term View* and never forget the implications that our decisions today have for the lives of millions.

Eric Goosby
*United States Global AIDS Coordinator*

Michel Kazatchkine
*Executive Director, The Global Fund to Fight AIDS, TB and Malaria*

Michel Sidibé
*Executive Director of the Joint United Nations Programme on HIV/AIDS (UNAIDS)*

# Executive summary

# A new approach for meeting the challenges of AIDS

The world is at a crossroads in the still-unfolding history of AIDS. In 30 years, an immense amount has been accomplished: scientific breakthroughs, unprecedented global funding, and a new model for human rights and public health policy. Most important, millions of lives have been saved. Despite these achievements, we are losing ground in the struggle: The response to the pandemic remains well behind the curve. Every day more than 7,000 individuals become newly infected with HIV—more than twice as many as the number of people who start on antiretroviral therapy each day. Although the continuing growth of the pandemic is sobering, we have the ability to bring it under control—with more and better science, smarter public policy, more efficient and effective programs, adequate funding, and strategies for addressing the blind spots in existing efforts.

aids2031 is an independent consortium of partners with experience in AIDS research, policymaking, and program design and practice, as well as expertise and perspectives from diverse fields, including economic, biomedicine, the social sciences, international development, and community activism. The consortium's mandate was to question conventional wisdom, stimulate new research, spark public debate, and examine social and political trends regarding AIDS. Initiated by UNAIDS in 2007, aids2031 has convened nine working groups that focused on modeling, social drivers, programmatic response, leadership, financing, science and technology, communications, hyperendemic countries, and countries in rapid economic transition. The consortium's charge was to bring new thinking to address the dilemma of a pandemic that is still growing despite great investments and efforts at control. Its focus has been to look at what should be done differently *now* to radically reduce the numbers of infections and deaths by 2031, the year that will mark 50 years since AIDS was first reported.

Tough choices have to be made in order to achieve a significant reduction in AIDS over the long term. *AIDS: Taking a Long-Term View* offers a vision of how our response to the pandemic can be reconceived to have more positive and sustainable outcomes, not only in the short term, but over the coming decades.

## Possible futures of the pandemic: opportunities and challenges

Much about the future of AIDS is known. The pandemic is certain to remain an extraordinary global challenge over the next generation and a leading cause of illness and death.

Yet much about the future of AIDS is currently unknowable. It is uncertain how the pandemic will evolve in its diverse settings and how effective the responses to it will be. In the coming years, changing global social, political, technological, and economic context will continue to affect both the pandemic and our efforts to respond. Some of these changes are predictable; others will take us by surprise. Increased economic development and prosperity may help sustain a renewed AIDS effort; new technologies and drugs will likely improve available diagnostics, prevention measures, and treatment.

However, other changes may create severe challenges. Further globalization and climate change are likely to increase migration and, hence, the spread of the disease. The world is generating the largest ever cohort of sexually active people in history, greatly expanding the numbers susceptible to HIV infections. A resurgence in risk behaviors is already apparent in some regions, including the world's richest countries, and new combinations of risk factors are occurring in some of the world's poorest nations and emerging economies. Progress remains meager in tackling the sexual and gender-based violence, exploitation, and discrimination that render women and girls disproportionately vulnerable to infection. Political instability and economic failures may create added pressures and distract attention from efforts to control the pandemic. A projected world population of more than 8 billion people by 2031, with large dependent populations of young people in some countries and elderly in others, will greatly increase the stress on funding for health. Population growth will also be a significant contributing factor to future numbers of people living with HIV, even if transmission rates decline.

There is of course, the possibility of a game-changer—the discovery of an effective vaccine or even a cure for HIV infection. Without either of these, however, we will probably never be able to eliminate AIDS. But with redesigned strategies to optimize currently available tools and support the introduction of new ones, it is possible to reduce the number of new infections to levels well below the current ones.

## The vision of aids2031

aids2031 asserts that a focus on long-term success has important implications for both future planning and current choices.

**New knowledge must be continually generated and used** in biomedical research as well as in more rigorous, field-based evaluation and learning. Research must focus on both effectiveness and efficacy. We will need to be vigilant to changes and adapt strategies, programs, and policies to stay relevant to the constantly evolving epidemics.

**Decision-makers must move from lip service to meaningful action on HIV prevention and prioritize it as the mainstay of a sustainable response.** The goal for prevention policies and programs should be to maximize the number of infections prevented. Bold leadership is needed to avoid policies and practices that stigmatize and marginalize groups or individuals at high risk of infection. Prevention efforts need to be more specifically focused on the populations and settings where they are most needed.

**An exceptional response will continue to be needed in Southern Africa.** Political leadership and accountability are sorely needed along with bold efforts to address the social drivers of HIV transmission. An all-out prevention campaign is needed and should include targeted strategies for young women and marginalized groups such as migrant labor, men who have sex with men, and injecting drug users.

**aids2031 proposes a minimum legal framework** for all countries which includes decriminalization of HIV status and transmission, sex work, and same-sex relationships and practices; ending barriers to the provision of harm reduction services for injecting drug users; destigmatization and equal rights for people living with HIV; as well as equality under the law for men and women.

**Additional strides are needed to improve treatment regimens and ensure their availability to all people living with HIV.** Historic achievements in expanding treatment access must not blind us to the reality that the current treatment model is not sustainable. In spite of marked declines in drug prices, standard antiretroviral regimens remain too expensive and complex to make life-long therapy feasible for tens of millions of individuals in the most resource-limited settings. In moving forward, the global community should adopt as its first priority extending life for the greatest number of people.

**While continuing to mobilize resources, improving efficiency will be critical** to reaching more people with needed prevention and treatment services. Synergies between prevention and treatment must be maximized.

**Nothing can replace the importance of leadership** in addressing the future. AIDS leaders have long criticized punitive laws and policies that impede a sound response to the epidemic and urged that funding targets those who need services the most. Less often have the political leaders who adopt punitive policies or ignore most-at-risk populations been called to account. That must change. International bodies, civil society groups, the news media, and other stakeholders need to be more willing to criticize those who undermine or obstruct effective action on AIDS. Structured reviews of national and subnational plans must be implemented to monitor progress toward adopting long-term strategies, plans, and budgets. Funders should significantly increase investment in independent civil society watchdog groups to monitor governments and other key stakeholders. Long-term sustainable change needs sufficiently robust budget lines with longer-term horizons of 10–20 years.

*AIDS: Taking a Long-Term View* **urges constant consideration of the long-term implications of our choices.** It recognizes that the choices that need to be made may not be politically popular, especially given that they may not show immediate success. But significant long-term change requires long-term thinking, long-term planning horizons, and long-term financing.

Much remains to be accomplished if future generations are to live in a world in which the threat of AIDS has been overcome. Social change does not come quickly or easily. The population of people living with HIV can contribute much to these efforts; a younger generation is socially conscious and looking for both opportunities and inspiration. A promise of positive synergies exists between the global AIDS response and responses to other diseases. Globalization also has brought new opportunities for communication, technical development, and supportive interactions to build new coalitions.

As with any effort that must be sustained over many years, complacency and denial are the enemies of long-term success. If we are to transform the landscape of AIDS by 2031, AIDS must remain high on the global agenda. We must move to a response that is long term and sustainable—one that makes full use of the knowledge and resources developed over the past three decades, yet continues to evolve and respond to a changing world that is constantly influencing the future of AIDS.

# AIDS timeline

1981 • Centers for Disease Control (CDC) publishes first report of AIDS

1982 • First community-based AIDS service provider created

   • First AIDS case reported in Africa

1983 • French scientists isolate HIV

   • Heterosexual transmission confirmed

   • Epidemic in central Africa documented

   • People With AIDS movement launched with Denver Principles

1984 • Jonathan Mann launches Projet SIDA in Zaire

1985 • First International AIDS Conference held

   • First test to diagnose HIV licensed

   • National AIDS response launched in Uganda

1986 • AIDS control program established at World Health Organization (WHO)

1987 • The AIDS Support Organisation (TASO) founded in Uganda

   • AIDS Coalition to Unleash Power (ACTUP) created

   • FDA approves azidothymidine (AZT) for treatment of HIV

1988 • The number of women living with HIV in sub-Saharan Africa found to exceed that of men

• First World AIDS Day held

1990 • Michael Merson appointed as Director of the WHO Global Programme on AIDS

• Less than one decade after the epidemic was recognized, an estimated 7.3 million people are living with HIV

1991 • Red ribbon first used as the international symbol of AIDS awareness

• International Council of AIDS Service Organizations formed

• Thai AIDS program moves from the Ministry of Public Health to the prime minister's office, initiating dramatic increases in AIDS funding

1992 • International Community of Women Living with HIV founded

• Rapid HIV test licensed

1993 • Results of Concorde study fail to demonstrate that AZT monotherapy extends life

• Regulatory authorities approve female condom

1994 • 42 countries formally adopt a declaration calling for Greater Involvement of People Living with HIV/AIDS

• Bill & Melinda Gates Foundation formed

• Report of an outbreak of HIV among paid blood donors in China

1995 • More than 18 million people worldwide are living with HIV

1996 • The Joint United Nations Programme on HIV/AIDS (UNAIDS) launched

   • Studies demonstrate that combination antiretroviral therapy significantly reduces risk of HIV-related illness and death

   • Free distribution of antiretroviral therapy begins in Brazil

   • International AIDS Vaccine Initiative (IAVI) launched

   • Number of new HIV infections—more than 3 million—is the highest in the epidemic's history

1997 • Global Business Council on HIV/AIDS launched in Edinburgh, U.K.

   • UNAIDS launches HIV Drug Access Initiative in Uganda and Côte d'Ivoire, representing the first introduction of antiretroviral therapy in sub-Saharan Africa

1998 • Treatment Action Campaign (TAC) forms in South Africa

1999 • The worldwide pace of AIDS deaths is rapidly increasing, with more than 1.5 million people dying each year

2000 • United Nations Security Council convenes first session on AIDS

   • Millennium Development Goals call for reversing the HIV epidemic by 2015

   • International AIDS Conference in Durban, South Africa, significantly quickens global momentum to expand HIV treatment access

   • World Bank launches Multi-Country HIV/AIDS Program (MAP)

- UNAIDS launches Accelerating Access Initiative (AAI)
- UNAIDS and WHO announce joint agreement with five pharmaceutical companies to lower prices for antiretrovirals
- The number of people living with HIV reaches 30 million

2001
- UN Secretary-General Kofi Annan calls for global "war chest" to fight AIDS
- African leaders gather in Abuja, Nigeria, for a historic summit on AIDS, committing to take specific action to strengthen national AIDS responses
- UN General Assembly's first-ever special session on HIV/AIDS results in adoption of time-bound pledges to strengthen AIDS response
- Doha Agreement allows developing countries to buy or manufacture generic drugs for HIV and other priority diseases
- South African President Thabo Mbeki appoints advisory body that includes AIDS denialists; the group issues a report recommending alternative therapies for AIDS
- Pretoria High Court orders government of South Africa to roll out a program to target the prevention of mother-to-child HIV transmission
- Pan Caribbean Partnership Against HIV and AIDS (PANCAP), a major regional initiative on AIDS, launched in Barbados

2002
- The Global Fund to Fight AIDS, Tuberculosis and Malaria launched

- AIDS becomes leading cause of death worldwide for people ages 15–59

2003 • U.S. President George W. Bush launches President's Emergency Plan for AIDS Relief (PEPFAR) initiative

- UNAIDS and WHO launch the "3 by 5" campaign to reach three million people living with HIV/AIDS in low- and middle-income countries with antiretroviral treatment (ART) by the end of 2005

- Clinton Foundation secures major price reductions for AIDS drugs

- First large-scale HIV vaccine trial announces results, with no efficacy found

- Avahan India AIDS Initiative launched to provide HIV prevention services to most-at-risk populations in high-prevalence states

- Wen Jiabao becomes first Chinese premier to shake the hand of an HIV-positive person

2004 • G8 nations call for creation of Global HIV Vaccine Enterprise

- More AIDS deaths—more than 2 million—occur than in any prior year

2005 • French researchers report that adult male circumcision reduces risk of female-to-male sexual transmission by 60%

2006 • Global community endorses goal of universal access to HIV prevention, treatment, care, and support by 2010

- UNITAID launched to create international drug purchase facility

2007  • New international guidelines issued for
        provider-initiated HIV testing in health care settings

      • Efficacy trial for most promising AIDS vaccine
        candidate halted due to lack of efficacy

      • International recommendations for adult male
        circumcision issued after two additional studies
        confirm 2005 study findings

      • UNAIDS and WHO announce that global HIV
        incidence appears to have peaked in mid- to late
        1990s

2008  • Coverage for antiretroviral therapy and services to
        prevent mother-to-child transmission exceeds 40%
        for the first time

      • Global financial and economic crisis emerges,
        threatening future financing for HIV programs

2009  • No efficacy found for most promising early-
        generation microbicide candidate

      • Surveys commissioned by UNAIDS suggest that the
        global financial and economic crisis is negatively
        affecting AIDS programs in many developing
        countries

      • Modest funding increase for PEPFAR falls well short
        of amounts authorized

2010  • South African President Jacob Zuma commits to
        dramatically strengthen the country's HIV
        prevention and treatment programs

      • Several African countries begin capping enrollment
        in antiretroviral treatment programs as a result of
        funding shortfalls

# List of acronyms

| | |
|---|---|
| **ACTUP** | AIDS Coalition to Unleash Power |
| **AIDS** | Acquired immune deficiency syndrome |
| **ARV** | Antiretroviral |
| **ART** | Antiretroviral therapy |
| **AZT** | Azidothymidine (antiretroviral drug) |
| **CAPRISA** | Centre for the AIDS Programme of Research in South Africa |
| **CDC** | Centers for Disease Control and Prevention |
| **CERN** | European Organization for Nuclear Research |
| **CHAI** | Clinton Health Access Initiative |
| **DHS** | Demographic Health Survey |
| **FDA** | Food and Drug Administration |
| **GDP** | Gross domestic product |
| **GIPA** | Greater Involvement of People with HIV/AIDS |
| **HAART** | Highly active antiretroviral therapy |
| **HIV** | Human immunodeficiency virus |
| **IAVI** | International AIDS Vaccine Initiative |
| **IMF** | International Monetary Fund |
| **IPCC** | Intergovernmental Panel on Climate Change |
| **IPPF** | International Planned Parenthood Federation |
| **IPM** | International Partnership for Microbicides |
| **NERCHA** | National Emergency Response Committee on HIV/AIDS (Swaziland) |

| **NGO** | Nongovernmental organization |
| **ODA** | Overseas Development Assistance |
| **PEPFAR** | President's Emergency Plan for AIDS Relief |
| **PrEP** | Pre-exposure prophylaxis |
| **R&D** | Research and development |
| **SMS** | Short Message Service |
| **STI** | Sexually transmitted infection |
| **TASO** | The AIDS Support Organisation |
| **TAC** | Treatment Action Campaign |
| **TB** | Tuberculosis |
| **UN** | United Nations |
| **UNAIDS** | The Joint United Nations Programme on HIV/AIDS |
| **UNDP** | United Nations Development Programme |
| **UNFCC** | United Nations Framework Convention on Climate Change |
| **UNITAID** | International facility for the purchase of drugs against HIV/AIDS, malaria, and tuberculosis |
| **UNODC** | United Nations Office on Drugs and Crime |
| **WHO** | World Health Organization |
| **WHOSIS** | World Health Organization Statistical Information Service |

# 1

## The future of AIDS: a still-unfolding global challenge

The world is approaching a moment of truth in the still-unfolding response to the AIDS epidemic.

The worldwide mobilization to combat a disease unheard of 30 years ago has generated historic achievements. For the first time, complex, lifelong management of a chronic disease has been widely implemented in low-income countries, averting millions of deaths.[1] Prevention services have also been introduced in antenatal settings to prevent infants from becoming infected, and the overall number of new HIV infections in 2009 for both children and adults was more than 20% lower worldwide than in 1997.[2]

The pandemic has also elicited an unprecedented global mobilization of political and financial resources. For the first time in history, a United Nations program was established dedicated to fighting a single disease and the first-ever special session of the UN General Assembly was held devoted to a particular health condition. AIDS also led to the creation of a major new addition to global health financing architecture, The Global Fund to Fight AIDS, Tuberculosis and Malaria. As of 2008, total annual resources for HIV programs in low- and middle-income countries reached US$ 15.6 billion, an astonishing 53-fold rise in 12 years.[3, 4] And, by 2010 an estimated 5 million people in low- and middle-income countries were receiving antiretroviral treatment, a remarkable 12-fold increase in less than a decade.[5]

Against this historic progress, ominous signs suggest that the AIDS response is beginning to fracture. In May 2010, the *New York Times* profiled Uganda's biggest AIDS clinic, where stalled funding has prompted authorities to cap the number of patients enrolled in HIV treatment.[6]

Citing Uganda as the "first and most obvious example of how the war on global AIDS is falling apart," the *Times* reported that the "golden window" of worldwide generosity appears to be closing, consigning countless newly diagnosed Ugandans to an early death.

The world has traveled this road before. From the Green Revolution that made acute hunger a thing of the past in scores of countries throughout the world to major initiatives that eradicated malaria from large swathes of the world, the global community has repeatedly mobilized to address global health inequities, only to lose interest or to declare victory prematurely. The results have been catastrophic, especially in Africa, where conditions that have been unknown in upper- and middle-income countries for decades continue to cost millions of lives each year.

Will this same tragic story be repeated with AIDS? Or will decisionmakers throughout the world learn from the mistakes of the past and chart a different, healthier, more enlightened course?

In 2031, the world will mark 50 years since AIDS was first reported. Recognizing the generations-long challenge posed by AIDS, UNAIDS launched aids2031 as an independent consortium, composed of experts in public health, economics, biomedicine, the social sciences, international development, and community activism. Organized into nine thematic working groups, the aids2031 Consortium has examined possible futures of the pandemic, the factors most likely to determine its future course, and the steps needed to sharply reduce the number of new HIV infections and AIDS deaths over the next generation.

## The aids2031 Consortium

aids2031 is a consortium of partners experienced in AIDS research, policymaking, and programming, as well as additional voices who have contributed expertise and perspectives from such diverse fields as economics, biomedicine, the social sciences, international development, and community activism. Conceived in 2006, 25 years after AIDS was first reported, and launched at the World Economic Forum's 2007 Annual Meeting in Davos, Switzerland, the consortium set out to bring new thinking to address the dilemma of a pandemic that is still growing despite great investments and efforts to control it. The consortium's answer to this charge is to offer recommendations for a transition from what has been a largely short-term response to one that is governed by a long-term perspective. The consortium focused on what needs to be done now to radically reduce the numbers of infections and deaths by 2031.

aids2031 has convened nine multidisciplinary working groups that include economists, epidemiologists, and biomedical, social, and political scientists to question conventional wisdom, stimulate new research, spark public debate, and examine social and political trends regarding AIDS. The nine Working Groups are hosted by different institutions and focus on modeling, social drivers, programs, leadership, financing, science and technology, communications, hyperendemic areas, and countries in rapid economic transition. Together the working groups have engaged more than 500 individuals around the world. Each working group has taken a different approach in generating analyses, from community dialogues in West and Southern Africa, to mathematical modeling of possible epidemic and financial futures, forums on trends in scientific discovery, global political stakeholder analyses, and youth leadership forums.

*AIDS: Taking a Long-Term View* examines the AIDS challenge, using a future-oriented lens to identify successful strategies that need to be strengthened and ways in which the response to AIDS must change. It directly refutes the growing "AIDS fatigue" reported among key international donors, some national governments, and global opinion leaders, and rejects the either/or choice between focused initiatives to address specific diseases and strengthened broad-based health services. Both approaches are needed.

If the AIDS landscape is to undergo transformation by 2031, when the world marks the 50th anniversary of the initial recognition of the pandemic, AIDS *must* remain high on the global agenda. A long-term and sustainable response is needed that continues to generate and apply new knowledge.

## Reflecting on the past, looking toward the future

The emergence of the epidemic more than a generation ago represented a historic and unexpected development. By the early 1980s, experts believed that the era of infectious disease was fast becoming a relic of an earlier era.

While developing countries might continue to grapple with old diseases that had been largely banished from industrialized settings but never brought under control globally—including malaria and tuberculosis—it was assumed that the world overall was entering an epoch when chronic diseases associated with rising affluence would consume the efforts of health planners and medical researchers.

AIDS upset all these expectations and assumptions. In three decades, HIV has infected more than 60 million people worldwide.[7] More than 27 million have died of AIDS-related causes. AIDS is the leading cause of death in sub-Saharan Africa and one of the leading causes of death worldwide among reproductive-age women.

The epidemic has inflicted the "single greatest reversal in human development" in modern history.[8] In sub-Saharan Africa, home to two-thirds of all people living with HIV, average life expectancy has fallen by more than a decade during the last 20 years as a result of AIDS (see Figure 1.1).

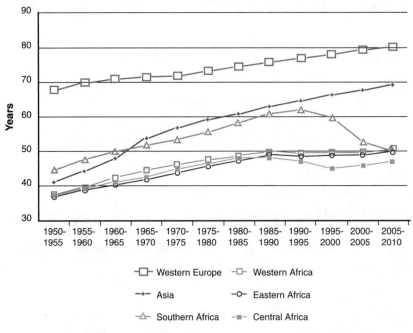

Source: Population Division of the Department of Economics and Social Affairs of the United Nations Secretariat/World Population Prospects.

Figure 1.1 Life expectancy at birth, selected regions 1950–1955 to 2005–2010.

Although sub-Saharan Africa has been most affected, other regions have not been spared. Even though HIV prevalence in Asia is only a small fraction of that in Africa, the region experienced 300,000 AIDS deaths in 2009.[9] Moreover, as Figure 1.2 illustrates, existing trends are not encouraging in Asia and the modes of transmission are changing. The pandemic costs affected Asian households more than US$ 2 billion annually and is projected to cause an additional 6 million Asian households to fall into poverty by 2015.[10]

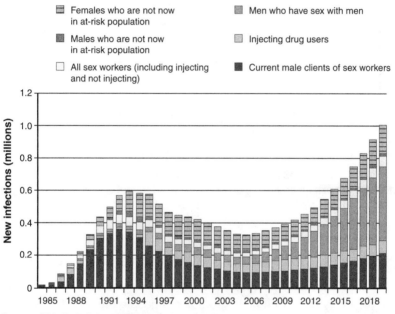

- ▦ Females who are not now in at-risk population
- ■ Men who have sex with men
- ■ Males who are not now in at-risk population
- ▢ Injecting drug users
- ☐ All sex workers (including injecting and not injecting)
- ■ Current male clients of sex workers

*Source: CAA: Redefining AIDS in Asia Technical Annex.*

Figure 1.2    Past and projected new HIV infections in Asia by population group.

Much is certain about the epidemic's future. AIDS will remain an enormous global challenge. Even with a robust and much stronger effort to prevent new infections and deliver effective therapies, the disease will undoubtedly remain a major cause of death worldwide. In Southern Africa, AIDS will continue to pose an existential threat to national economies, agricultural sectors, and both urban and rural communities.

Yet much about the future of AIDS remains uncertain. In large measure, the pandemic's severity in 2031 will depend on choices to be made in the next few years. This chapter summarizes the sobering results of modeling undertaken by the aids2031 Modeling Working Group. If efforts to tackle AIDS become smarter, more focused, and more community centered, tens of millions of lives can be saved over the next generation. If actions to address AIDS instead remain static or weaken over time, the result will be millions of preventable new infections and AIDS will claim many more millions of lives.

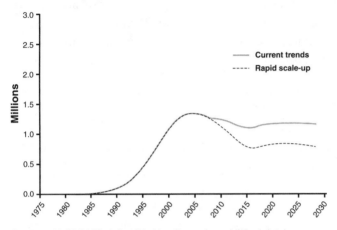

Source: aids2031 Modeling Working Group (unpublished data)

Figure 1.3   Annual AIDS deaths (adults 15–49 years) comparing current trends against expanded scenario. The cumulative number of deaths avoided between 2008 and 2031 in the expanded scenario is more than 7 million.

## What the history of AIDS may tell us about the future

More than 33.3 million people worldwide are living with HIV, and nearly 5,000 die every day. Each hour, around 200 people die of AIDS. There are around 7,400 new infections every day.[11] The statistics are numbing, conveying the extraordinary human toll of the disease.

What was once a new disease has long since become familiar. The well-documented history of the epidemic tells us important facts that are relevant to the future. Consider some of the known facts.

### Epidemics often differ radically within and between countries and regions

The so-called "global AIDS epidemic" is, in reality, an amalgamation of multiple local epidemics that often differ markedly from one another. Although women account for 60% or more of all people living with HIV in sub-Saharan Africa, men tend to predominate among HIV cases in most other regions.[12] Whereas sexual intercourse is the primary mode of transmission in India overall, extremely high rates of infection are reported among people who inject drugs in certain districts.[13] In Benin, Kenya, and Tanzania, the variation between the highest-prevalence district and the lowest is 12-fold, 15-fold, and 16-fold, respectively.[14] Meanwhile, HIV prevalence in Côte d'Ivoire is more than twice as high as in Liberia or Guinea, even though these countries share national borders.[15]

These and other variations teach us is that AIDS programs and policies must address the unique set of circumstances in particular settings. Certain principles may apply to AIDS responses everywhere—such as the value of a rights-based approach or the importance of engaging affected communities in the response—but no cookie-cutter model exists for addressing the broadly divergent types of epidemics around the world.

### The pandemic is constantly evolving

Even when epidemiological trends appear stable, the pandemic is constantly changing. As Figures 1.4–1.6 illustrate, what began as epidemics with differing characteristics primarily confined to a handful of countries progressively spread to affect the entire world.

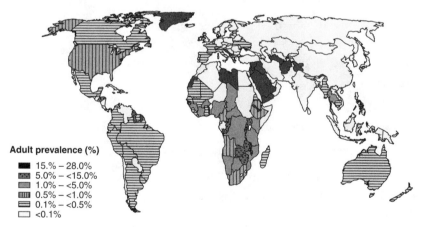

Adult prevalence (%)

■ 15.% – 28.0%
▨ 5.0% – <15.0%
▤ 1.0% – <5.0%
▥ 0.5% – <1.0%
▣ 0.1% – <0.5%
☐ <0.1%

Source: Adapted from UNAIDS data.

Figure 1.4   Progress of the AIDS pandemic, 1990.

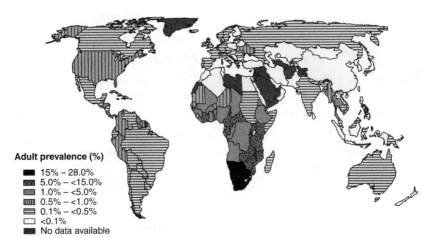

Source: Adapted from UNAIDS data.

Figure 1.5    Progress of the AIDS pandemic, 2000.

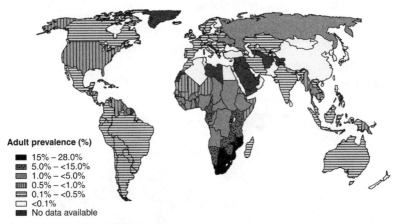

Source: Adapted from UNAIDS data.

Figure 1.6    Progress of the AIDS pandemic, 2009.

Within regions and countries, the nature of individual epidemics has changed over time. In China, Eastern Europe, and Central Asia, epidemics that were previously characterized primarily by transmission via injecting drug use are now increasingly driven by sexual transmission.[16] In many European countries, epidemics that were earlier concentrated in gay communities have given way to epidemics in which heterosexual adults and immigrants to the region are also at risk of infection.[17] As the epidemic has matured and become endemic in sub-Saharan Africa, older adults in stable, long-term relationships now account for the largest share of new infections in many African countries.[18]

*The evolution of each epidemic is affected by its social, economic, and physical environment*

Patterns of social and economic life often determine the trajectory of local epidemics. In New York City, a "synergism of plagues," abetted by the planned shrinkage of municipal services as a result of the city's acute fiscal crisis in the 1970s, was directly tied to the explosion of HIV in the early 1980s among low-income drug users.[19] The astonishing emergence of South Africa's HIV epidemic in the 1990s—with estimated HIV prevalence rising from less than 1% at the beginning of the decade to nearly 20% by the turn of the century—coincided with the dramatic social and population dislocations associated with the demise of apartheid. In Eastern Europe and Central Asia, the disintegration of the Soviet Union triggered radical shifts in sexual and drug-using behaviors and the rapid deterioration of public health services, contributing to sharp increases in HIV transmission.

The history of human relations and labor patterns in Southern Africa is indelibly tied to the epidemic's severity in that subregion. The countries with exceptionally high HIV prevalence are generally called "hyperendemic." Southern Africa is home to nine hyperendemic countries where adult HIV prevalence exceeds 10%, about which the aids2031 Hyperendemic Working Group concludes: "A central historical feature shared by all the hyperendemic countries was the rapid, forced proletarianization of males, the establishment of circular migratory patterns and, post-independence, large-scale urbanization."[20] These trends triggered broad-scale migration of male workers, disrupted households, contributed to high rates of sexual concurrency, and destroyed traditional methods for setting social norms. The eventual results are evident in South Africa's rapidly growing informal urban settlements, where infection rates are twice the national average.[21]

*The epidemic has become firmly entrenched in Southern Africa*

AIDS is a pressing health challenge for scores of countries, but its effects are especially pronounced in Southern Africa. With just 2% of the world's population, Southern Africa accounts for 34% of all people living with HIV.[22] It is impossible to speak of the future of Southern Africa without discussing the future of AIDS.

As Figure 1.7 reveals, the epidemic has skewed population structures in countries such as Lesotho. The figure for Ghana shows a more typical change in age structure resulting from lowered birth rates and improved life expectancy. By contrast, Lesotho, with one of the world's highest levels of HIV infection, shows a marked depletion of the population of working-age adults. In hyperendemic settings, the extraordinary loss of adults in their 30s and 40s has interrupted the natural process that imparts learning and values to younger generations, resulting in national populations that consist largely of the very young and the very old. These patterns not only reflect the epidemic's extraordinary impact, but also illustrate the degree to which AIDS undermines the ability of hyperendemic countries to mount a robust and sustained response to the epidemic.

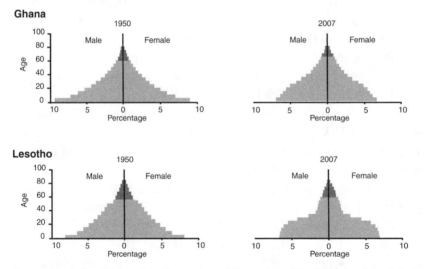

*Source: Population Division of the Department of Economic and Social Affairs of the United Nations Secretariat, World Population Prospects: The 2006 Revision.*

Figure 1.7   Changes in population structure: Ghana and Lesotho.

## Swaziland: the worst AIDS epidemic in the world

The little country of Swaziland, home to about a million people, has the dubious distinction of having the world's worst national AIDS epidemic. Although Swaziland is unmatched in the seriousness of its national epidemic, the AIDS challenge is comparable to other subnational areas, such as KwaZulu-Natal province in South Africa.

The earliest AIDS case in Swaziland was in 1986. In 1992, the first sentinel survey of antenatal clinic attendees was conducted, revealing a prevalence of 3.9%. Biannual surveys have subsequently tracked the exponential spread of the virus (see Figure 1.8). By 2004, Swaziland had the highest prevalence ever recorded. The small decrease in HIV prevalence in antenatal settings in 2006 was reversed in 2008, although the reversal may be indicative of increased numbers of women accessing treatment, placing upward pressure on HIV prevalence by reducing the rate of AIDS deaths.

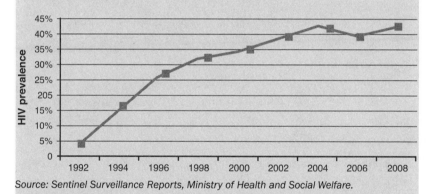

Source: Sentinel Surveillance Reports, Ministry of Health and Social Welfare.

Figure 1.8   HIV prevalence among antenatal clinic attendees in Swaziland, 1992–2008.

In 2008, the Demographic and Health Survey (DHS) found that 18.8% of the national population was living with HIV. Figure 1.9 shows this by age and gender. Note that adult HIV prevalence is considerably higher than national prevalence: 26.1% in 2007, according to UNAIDS. Prevalence is higher among women than men, and highest in 25- to 29-year-olds.

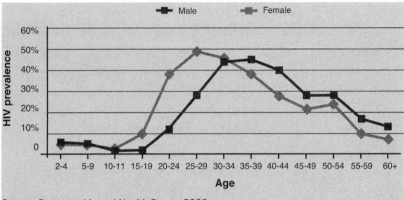

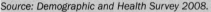

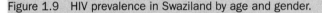

Source: Demographic and Health Survey 2008.

Figure 1.9   HIV prevalence in Swaziland by age and gender.

## Drivers of the epidemic

Why is the epidemic so serious in Swaziland? What has driven prevalence? If the answer were known, interventions would be clear. Studies and speculation have not yielded a single neat answer.[23] Instead, Swaziland's unmatched epidemic can most plausibly be traced to a unique combination of biological, economic, social, and cultural factors.

Migration has been common for decades and increases risk. For example, many Swazi men once worked in South Africa's mining industry. One study on a sugar plantation found that 24% of daily commuters and 41% of monthly commuters were infected. Permanent workers had a prevalence of 33.7%, casual workers 48.6 %.[24]

The epidemic is also perpetuated by multiple concurrent partnering and intergenerational sex in which girls and young women have sex with older men (see Figure 1.9). No evidence suggests that Swazis have more partners, or more frequent sex, than other groups in southern Africa. However, polygamy is culturally accepted. The Demographic Health Survey shows HIV prevalence is higher for men in polygamous marriages than for those in monogamous unions (47% and 30%, respectively). However, few Swazis are in formal polygamous relationships (2% of men and 7% of women), so of greater concern is the message that polygamy communicates about male sexuality, gender equity, and female status.[25] Although a national male circumcision policy was launched in August 2009, lack of male circumcision still contributes to the spread of HIV in Swaziland.

Sexual and gender-based violence disproportionately affect women and girls of all ages and are major risk factors. Sexual violence against females occurs across all socioeconomic and cultural backgrounds; many societies socialize women to accept, tolerate, and rationalize such experiences and, worse, to remain silent about them.[26] Among 18- to 24-year-old females, nearly 66% have experienced sexual violence in Swaziland. Overall, 48% of females report experiencing some form of sexual violence in their lifetime.

## Responses

Early responses to the epidemic in Swaziland were consistent with international best practice. By 1987, blood for transfusions was screened, information and education programs rolled out, condoms promoted, and a National Prevention and Control Program established.

The potential threat to the nation's social and economic well-being was discussed as early as 1992, and the Ministry of Economic Planning commissioned work on the socioeconomic impact of HIV/AIDS.[27] In 1999, King Mswati III declared AIDS a "national disaster" and the National Emergency Response Committee on HIV/AIDS (NERCHA) was established. Swaziland has successfully mobilized resources from donors, including the U.S. government and The Global Fund to Fight AIDS, Tuberculosis and Malaria. By 2009, 89% of adults needing treatment were on antiretrovirals, 59% of HIV-infected children were on treatment, and many orphaned and vulnerable children were being supported.

### Concept of an emergency

Swaziland's epidemic is so severe that it has led to a reassessment of the concept of an emergency.[28] The work used key socioeconomic indicators on demographic changes, health, social indicators, orphans, economic growth, investment, and agriculture to build a picture of the multidimensional impact of the disease.

The picture is bleak: Swaziland is experiencing a humanitarian crisis comparable to conflict countries or those facing a severe natural disaster. AIDS here is a slow-onset disaster, leading to a long-term catastrophe and requiring an urgent, sustained response. One challenge is the country's classification as "lower-middle-income," making it ineligible for some development assistance. An immediate response is needed while simultaneously building capacity for the long-term.

## AIDS discriminates

Although AIDS epidemics began among the most affluent in some countries, they almost invariably visit their harshest effects on marginalized
groups. Adult HIV prevalence for the world as a whole is 0.7%, yet an
estimated 13% of people who inject drugs, 6% of men who have sex with
men, and 3% of sex workers are living with HIV.[29] Moreover, these global
estimates, derived from national surveys, significantly understate the
pandemic's burden on these populations in particular settings. National
surveys in Malawi, for example, indicate that more than 70% of sex workers in the country are infected.[30] Even in sub-Saharan Africa, where heterosexual intercourse has long been assumed to be the almost exclusive
driver of epidemics, recent studies have documented exceptionally high
HIV prevalence among these key populations. As Figure 1.10 illustrates,
multiple studies have documented levels of HIV infection among men
who have sex with men that consistently exceed overall adult HIV prevalence in these settings. In Kenya, men who have sex with men, people
who inject drugs, and sex workers and their clients account for roughly
one in three new HIV infections.[31] In Senegal, men who have sex with
men are believed to represent up to 20% of all incident infections.[32]

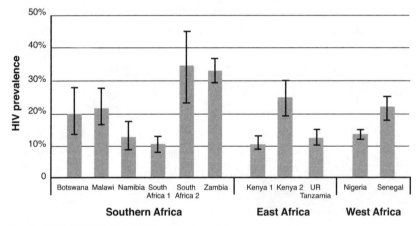

Source: WHO/UNAIDS 2009 AIDS epidemic update.

Figure 1.10   HIV prevalence among men who have sex with men in sub-
Saharan Africa (2002–2008). [Note: South Africa 1 and 2 and Kenya 1
and 2 refer to the results from different studies.]

*The pandemic's history cautions us to anticipate unexpected turns over the next generation*

Although HIV information systems remain weak in many countries (an issue that Chapter 2, "Generating knowledge for the future," addresses in some detail), the ability to monitor and understand national epidemics has greatly improved. After experiencing a rapid increase during its first 15 years, the pandemic appears to have stabilized globally, with the annual number of new infections about 20% lower today than in the mid-1990s.[33] With the exception of Eastern Europe and Central Asia, where new infections continue to increase, the pandemic appears to have stabilized or slowed in most regions.

This apparent stabilization of the pandemic has given rise to a general consensus in the popular media that the future of the pandemic can be projected with some accuracy. AIDS, it is said, has "peaked," with a slow but steady decline in rates of new infections likely to occur in the foreseeable future.

This emerging confidence in our ability to predict the future of the pandemic ignores previous experience with other infectious diseases and with AIDS itself.[34] Studies of endemic syphilis, for example, have documented that waves of infections are common over time, with spikes in incident cases often separated by a decade or more.[35] This is especially likely for epidemics that have become concentrated in particular populations. After a sharp decline in HIV incidence among men who have sex with men more than two decades ago in high-income countries, rates of new infections have steadily increased since the early 1990s.[36] In part, this reflects the fact that new cohorts of young people enter the population of sexually active adults over time.

Indeed, AIDS has repeatedly defied predictions and is sure to do so in the future. In December 1995, the WHO Global Programme on AIDS erroneously projected that the pandemic's center would be in Southeast Asia, with more modest growth predicted in sub-Saharan Africa.[37] A decade ago, few would have predicted that more than 1 million people would be living with HIV today in the Russian Federation. And certainly few observers foresaw a reversal in Uganda's longstanding gains in reducing HIV prevalence.[38] On the more favorable side of the ledger, only a small number of scientists were prepared for the emergence in the mid-1990s of combination antiretroviral therapy, which, for those who had access to it, rapidly converted the disease from an invariably fatal condition to one that is chronic and manageable.

The pandemic's past teaches that political, economic, and social shocks can greatly affect the trajectory of AIDS. In this regard, the accelerating rate of population migration in many regions is cause for concern. A study of six Asian countries by the aids2031 Working Group on Countries in Rapid Economic Transition concludes that major, continued population movements, particularly associated with growing urbanization, may facilitate the spread of HIV over the next generation.[39] In China alone, 300 million people are expected to migrate over the next 20 to 30 years.[40] Although mobility on its own is not a risk factor for HIV, migration often increases risk and vulnerability by disrupting social and familial networks, contributing to sexual risk taking and drug use, and subjecting individuals to violence and discrimination. In Asia, population migration has been strongly linked with the spread of infectious diseases.[41]

Other changes are also foreseeable. The introduction of antiretroviral therapy in resource-limited settings is rapidly altering attitudes about AIDS in many parts of the world where infection has long been regarded as a death sentence. Although the availability of treatment is arguably a prerequisite for a strong, sustainable effort to prevent new HIV infections (an issue Chapter 4, "Financing AIDS programs over the next generation," addresses in greater detail), the health benefits of treatment may also cause some to view the disease with less concern and alarm. In high-income countries, especially among men who have sex with men, evidence indicates that sharp reductions in HIV-related death and illness have contributed to an increase in sexual risk behaviors, ultimately resulting in rises in HIV incidence.[42] Were such trends to be replicated elsewhere, the results could be potentially catastrophic.

In short, much about the future of the pandemic remains unclear. This uncertainty necessitates continued vigilance in the AIDS effort worldwide.

## How will the virus itself evolve?

National epidemics evolve in response to changes that affect vulnerability to infection, but the virus itself is also evolving. In comparison to human evolution—which, from our vantage point, occurs quite slowly—the virus may evolve rapidly.[43]

One outstanding question is what evolutionary impact the still-recent introduction of highly active antiretroviral therapy will have on the evolution of HIV. For example, if viral load and HIV transmissibility are heritable traits, they may introduce new selective pressure.[44] One result may be a decline in overall HIV prevalence, but with prevalent infections more likely to be characterized by highly virulent, highly resistant, and easily transmissible virus. Thus, one possible future of AIDS could be that infection will become somewhat less common, but that the disease could be far more lethal to those who are infected. Another could be that the virus evolves to have less virulence, but affects more people.

## AIDS in a changing world

As a disease that is fundamentally linked to the way humans live and how they relate to one another, AIDS is inextricably entwined with the future of our world—and our world is rapidly changing.

### Globalization

Regions that once seemed remote from each other have drawn considerably closer as a result of increased international travel, breakthroughs in communications technology, and the internationalization of commerce, social trends, and political groupings. These trends have already had important effects on the pandemic and will continue to do so.

Globalization may bring benefits as well as challenges. Whereas humans have historically concerned themselves primarily with problems in their own countries or communities, the increasing inter-connectedness of our world makes it possible to mobilize global endeavors to address global problems. Thus, an Irish rock star can galvanize global attention on the pandemic's intense burdens in sub-Saharan Africa, and an African-born player for a major European football club can focus attention on problems in his home country.

A key driver of globalization, the revolution in information and communication technology, is changing the ways people communicate about behaviors and issues.[45] Although speaking about the "digital divide" between rich and poor countries is common, this gap is narrowing quickly, as use of the Internet and mobile communications technology is growing fastest in developing countries. Technological advances also may upend historic patterns; in developing countries, for example, women are more likely than men to use SMS text messaging to communicate. As with other aspects of globalization, communications advances may provide new avenues for intervention while at the same time raising new challenges. Social networking technologies offer new ways to mobilize communities and societies to take action, but they may also facilitate increased risk behavior; in some countries, sex work solicitation is rapidly transitioning from brothels, streets, and other traditional venues to the Web and mobile phones.

Globalization also teaches us something else. We sometimes speak as if we are living in a unique moment in human history, but this is not the first era of globalization. Between 1890 and 1913, levels of international trade and financial transactions were comparable to today's.[46] But unforeseen political and economic shocks, including World War I and the Great Depression, brought this earlier era of globalization to an abrupt end. This history reminds us that although we can—and should— do our best to anticipate future trends, unexpected surprises may well be in store, testing our ability to adapt to a radically different set of circumstances.

### Climate change

Extreme weather events and other disasters associated with climate change, as well as the increased frequency and severity of droughts in developing countries, are likely to generate up to 150 million climate-change refugees in coming decades[47] and further accelerate the exodus from rural villages to urban settings. Although mobility itself is not a risk factor for HIV, large-scale population dislocations frequently place people in situations of increased risk and vulnerability. By exacerbating the degradation of agricultural sectors, climate change may worsen the well-documented effects of AIDS on household food security and agricultural economies in sub-Saharan Africa.

Climate change could have other real, although indirect, effects on the future of AIDS by reducing funding for health programs in low- and middle-income countries. The United Nations Framework Convention on Climate Change projects that developing countries will incur annual costs of adapting to climate change of US$27 billion to 66 billion by 2030[48]; other researchers predict that associated costs will be considerably higher.[49] With such extraordinary costs looming, developing countries and external donors may struggle to accommodate other competing needs, such as AIDS, other infectious diseases, or health-systems strengthening.

## Climate change and AIDS: some similarities and one notable difference

AIDS and climate change are linked not only in their possible effects on human behavior and their potential competition for scarce resources. Both issues have also given rise to massive global movements.

AIDS and climate change are alike in another way. Just as AIDS produced its own chorus of "denialists," who resisted the overwhelming scientific evidence linking the disease to HIV, climate change has generated a minority of skeptics who disregard science and dispute the evidence of the human contribution to climate change.

However, the single largest contrast between the AIDS movement and the global environmental movement is perhaps more revealing. As the following chapters of this book explain, the response to AIDS has, from the very beginning, been framed as an emergency response. Since AIDS first appeared, time has been of the essence. Instead of building for the future to sustain a generations-long fight, the AIDS movement has tended to seek immediate results.

By contrast, the global effort to slow and reverse climate change has from its inception been understood as a long-term undertaking. Even the most ardent activists have understood that weaning the world off fossil fuels will take time. And the extensive planning focused on coping with the impact of climate change has expressly sought to achieve results only decades down the road.

### Population growth

By 2031, the global population is projected to exceed 8 billion people. Countries are already finding that stability in the percentage of the national population infected with HIV translates over time into increasing numbers of people living with the disease. Population growth also has the potential to increase social conflict regarding natural resources such as water or food, which could give rise to greater population mobility and further increase risks of and vulnerabilities to HIV.


## A changing global power structure

Existing global structures and mechanisms for AIDS decision-making arose out of political and economic power structures put in place after World War II. The victors in that conflict forged global institutions that have played central roles in the AIDS response, including the United Nations system and the World Bank. The major AIDS donors also have largely reflected the Atlantic orientation of global power in the second half of the 20th century.

These power dynamics are rapidly changing. China, India, and other Asian economies are growing far more rapidly than those represented by the Group of Seven major industrialized countries. Brazil, Russia, South Africa, and other countries are also rapidly coming into their own as global and regional powers, illustrated most vividly by the G20's replacement of the G7 as a key forum for global decision-making. Meanwhile, the traditional global powers, most recently buffeted by the global financial and economic crisis, confront worrisome structural challenges associated with economic stagnation, long-term budget deficits, and rapidly aging populations.

How these trends will affect the future of AIDS is uncertain. The relative political and economic decline of the donors that have helped underwrite the massive build-up of financial resources for AIDS is a cause for concern, at least. And the transition to a more multipolar world could conceivably make it even more difficult to marshal coordinated global action on major problems. But these trends also offer potential opportunities, as the corresponding development of new global powers offers the prospect of new AIDS donors coming on the scene in future years.

## Scenarios for the pandemic's future

Perhaps the clearest message of the pandemic's history to date is that concerted action can make a difference. Antiretroviral treatment had saved 2.9 million lives worldwide between 1996 and 2008.[50] Similarly, as Figure 1.11 shows, expanded access to services to prevent mother-to-child HIV transmission averted at least 200,000 new infections globally between 1996 and 2008, with these life-saving results increasing from year to year as services have been expanded.[51]

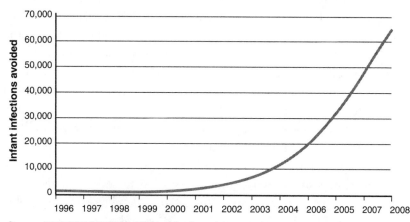

Source: WHO/UNAIDS 2009 AIDS epidemic update.

Figure 1.11   Estimate of the annual number of infant infections averted through the provision of antiretroviral prophylaxis to HIV-positive pregnant women globally (1996–2008).

The magnitude and quality of the continued response to AIDS will be determining factors in the pandemic's future. To clarify the long-term implications of choices to be made in the coming years, the aids2031 Modeling Working Group produced a series of mathematical models on potential AIDS futures in 22 countries, including 12 in Africa (Cameroon, Ethiopia, Kenya, Malawi, Mozambique, Nigeria, South Africa, Sudan, Tanzania, Uganda, Zimbabwe, and Zambia), six in Asia (Cambodia, China, India, Indonesia, Thailand, and Vietnam), two in Eastern Europe (Russia and Ukraine), and two in Latin America (Brazil and Mexico). For each of these countries, the model took into account the interaction between different sexually transmitted diseases and HIV, the role of heterogeneity in sexual behavior, patterns of sexual acts within partnerships, networks of concurrent sexual partnerships, patterns of incidence as a function of age, and the impact of interventions. For intervention impact, the modelers relied on the public health literature, including clinical trials and epidemio- logical studies; the models all assumed constant coverage and impact from 2015 onward. Models further assumed that receipt of antiretroviral therapy would reduce the likelihood of onward HIV transmission, a conclusion that is consistent with numerous studies that have correlated viral load with transmissibility.

Using these various data sources, modelers charted the likely future of epidemics in these 22 countries according to various scenarios, calcu- lating the number of incident infections, total HIV prevalence in 2031, and the number of AIDS deaths likely to occur under each scenario between 2010 and 2031. Importantly, the models aim to quantify the long-term impact of choices that will be made in the next several years, tracing the ultimate impact in 2031 to political choices made between 2010 and 2015. One possible future scenario, the "status quo" scenario, calculates results based on a continuation of current coverage levels. A second scenario envisages an intensification of HIV prevention and treatment programs toward saturation coverage.

The results of the modeling exercises led to several fundamental conclusions about the epidemic's future and the choices facing global decision-makers over the course of the next several years.

### Choices made in the next five years will profoundly affect how the pandemic will look in 2031

Maintaining existing coverage levels would allow nearly 50 million cumu- lative new infections among 15- to 49-year-olds in these 22 countries by 2031. By contrast, as shown in Figure 1.12, expanded coverage would avert more than 26 million of these infections (or more than half).

The impact of these choices in specific countries is revealing. Consider South Africa, for example, home to the world's largest number of people living with HIV. If current coverage levels continued, HIV prevalence in 2031 would be the same (18%) as it was in 2008, the most recent year for which data is available.[52] Yet this stability in prevalence is deceiving; with the projected growth in population, a stable prevalence would mean 12 million new infections between now and 2031. By con- trast, if AIDS response efforts are strengthened over the next several years, HIV prevalence would be one-third lower in 2031, assuming that scaled-up treatment will help slow the rate of new infections by lowering the level of virus circulating in communities. In Nigeria, Mozambique, and Zambia, reductions would be even sharper, with projected prevalence in 2031 nearly 50% lower under an intensified response. The flattening of all the curves after 2015 reflects the assumption that all intervention coverage and impacts is constant after that date.

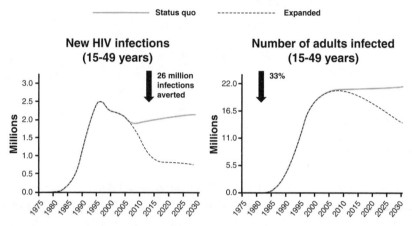

Source: aids2031 Modeling Working Group (unpublished data)

Figure 1.12    Projection of new HIV infections and number of adults infected according to status quo and an expanded scenario based on information received from 22 countries.

In Zambia, the annual number of new infections in 2031 would be nearly four times greater with a continuation of current service coverage than with an intensified, expanded response, as Figure 1.13 shows. In China, where population growth will be a critical driver in the number of new HIV infections over the next generation, the status quo scenario would result in more than three times as many incident infections in 2031 as in an expanded response, also shown in Figure 1.13. HIV-related deaths in Zambia would be more than twice as high in 2031 if current coverage levels continue, and the mortality rate in China would be more than three times as high.

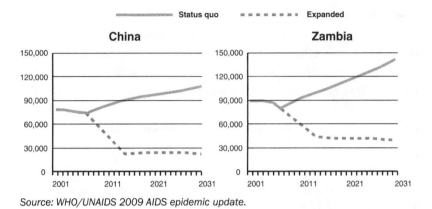

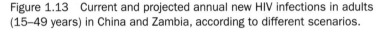

Source: WHO/UNAIDS 2009 AIDS epidemic update.

Figure 1.13   Current and projected annual new HIV infections in adults (15–49 years) in China and Zambia, according to different scenarios.

## To achieve dramatic change, all available tools must be used to their maximum advantage

To avert tens of millions of deaths over the next generation, as projected in the most favorable scenario, the best results possible must be obtained with the available tools. This demands continuous quality improvement, results-based management, and program evaluation to improve perform-ance over time (see Chapter 3, "Using knowledge for a better future"). Such improvements over time could be expected to continue beyond 2015, leading to better future results than shown in the projections.

## Prioritizing HIV prevention is critical to accelerated progress between now and 2031

The level of resources devoted to HIV treatment in low- and middle-income countries is roughly 2.5 times greater than amounts dedicated to HIV prevention.[53] Increasingly, AIDS efforts are characterized by an approach that focuses almost exclusively on treating existing infections and devotes meager resources to preventing new infections. This approach mimics the path high-income countries have taken. In the United States, for example, only 3% of government outlays for HIV are currently allocated to prevention programs.[54] Without substantial targeted HIV prevention efforts, new HIV infections will continue to outpace treatment efforts— even while recognizing some prevention efforts from expanded treatment.

Figure 1.14 demonstrates that the long-term results of this approach would be potentially devastating in resource-limited settings. In Zambia, nearly twice as many incident infections would occur in 2031 under a treatment-only approach as would occur with a combination of robust prevention *and* treatment efforts.

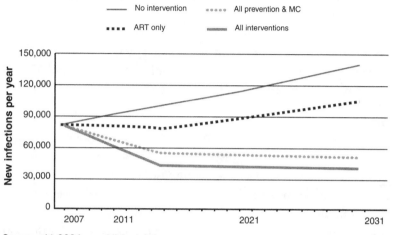

Source: aids2031 unpublished data.

Figure 1.14   Current and projected impact of intervention strategies on new HIV infections among adults in Zambia.

## To achieve optimal results for 2031, new prevention tools will be needed

To make truly revolutionary gains in the epidemic over the next two decades, new and better prevention tools are needed, along with structural changes in many communities and societies. According to the aids2031 Modeling Working Group, achieving a 90% reduction in HIV incidence by 2031 necessitates a 70% reduction in the average number of sexual partners.[55] This effect far exceeds results obtained with existing prevention tools, pointing to the need for additional prevention options and a broader approach to prevention that accounts for social drivers of national epidemics.

Figure 1.12 is telling. With an expanded response that achieves maximum coverage for both HIV prevention and treatment, the annual number of new HIV infections in Zambia would be half what it was in 2001 and roughly 50% lower than current HIV incidence. This would represent significant progress, but it would still leave Zambia grappling with an enormous health crisis that would sap national resources and push countless households into poverty.

The world should aim higher. However, reaching this lofty goal will require new scientific tools and improved knowledge about ways to change sexual behavior and prevent new infections, topics that the following chapter addresses in depth.

### Delivering treatment to those who need it will be vital to minimizing the pandemic's impact

Securing the gains envisaged in the intensified/expanded scenario will require continued, sustained support for scaling up HIV treatment. As Figure 1.15 illustrates, current coverage trends would cover fewer than half the number of people who would receive treatment under the optimal scenario in 2016. By 2031, the number of people receiving treatment in the optimal scenario would be more than 60% higher than with current trends.

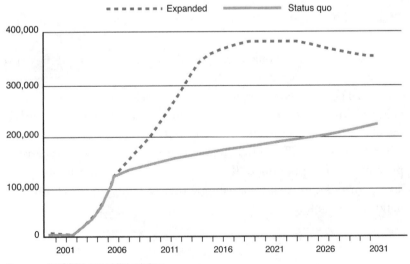

Source: aids2031 unpublished data.

Figure 1.15   Current and projected number of adults (15–49 years) receiving treatment in Zambia.

### These projections underscore the need to intensify measures to mitigate the pandemic's impact

An important lesson learned thus far in the global AIDS response is that even the most heavily affected societies have proven to be far more resilient than projected earlier in the epidemic.[56] Yet this resilience masks enormous individual and societal burdens, many of which are likely to endure for generations.

A case in point is the extraordinary number of the world's children who have been orphaned by the pandemic. In sub-Saharan Africa, more than 14.8 million children have lost one or both parents to AIDS.[57] As the projections summarize, vulnerable households will still face significant numbers of deaths and further burdens, even under the most favorable scenarios.[58]

Over the last three decades, the primary focus of the AIDS response has been on getting programs up and running, frequently in communities with little health infrastructure. In many cases, the goal has been to achieve immediate results. In some cases, responses have been premised on the expectation that an imminent biomedical breakthrough will resolve the need for further action. Typically, more difficult challenges— such as changing community norms or gender relations, or addressing the impact of labor or economic structures on individual or collective vulnerability—have been dismissed as too long range and time-intensive to merit investment in the context of an emergency.

The overarching theme of this report is that AIDS is a generations-long challenge and thus requires a generations-long response that adopts a longer-term mindset. In moving forward, the imperative of scaling up must be matched by an equally strong commitment to quality, efficiency, and sustainability. And, to achieve long-term success with AIDS, underlying drivers of national and local epidemics must be addressed, even if these efforts are unlikely to achieve results in the short term. Table 1.1 details the characteristics of the long-term and short-term approaches.

As with any effort that must be sustained over many years, complacency and denial are the enemies of long-term success; already, signs of the world's fading interest in AIDS are apparent. But with concerted efforts and a changed approach as described in this book, success in the AIDS response is achievable. A great deal of the knowledge needed to radically reduce the number of new HIV infections and AIDS deaths over the next generation is already available, and the world possesses the research capacity to generate the new preventive and therapeutic tools that will be required. Even in the midst of worldwide economic uncertainty and anxiety, little doubt arises that sufficient resources exist to address AIDS and other global health challenges.

The subsequent chapters detail the steps needed to place a global AIDS response on a long-term and sustainable footing. In the final chapter, the aids2031 Consortium offers recommendations for the steps that need to be taken now to ensure long-range success.

**Table 1.1  Selected characteristics of shifting from a short-term to long-term approach**

| Short Term | Long Term |
|---|---|
| Reactive | Proactive |
| Generic approach | Locally adapted approaches |
| Donor-imposed short-term (1–2 year) funding | Locally designed multiyear plans and financing (5–15 years) |
| External consultants | Investment in local capacity |
| AIDS isolated from broader health and development | Synergies with other health and development |
| Individual behavior change | Societal and individual behavior change |

## Endnotes

1. UNAIDS and WHO, *AIDS Epidemic Update* (Geneva: UNAIDS and WHO, 2009).

2. UNAIDS, *AIDS Info: UNAIDS Reference Report* (Geneva: 2010).

3. UNAIDS and the Kaiser Family Foundation, *Financing the Response to AIDS in Low- and Middle-Income Countries: International Assistance from the G8, European Commission, and Other Donor Governments in 2009*, July 2010. Accessed 2 August 2010 at http://www.kff.org/hivaids/7347.cfm.

4. UNAIDS, *Report on the Global HIV Epidemic* (Geneva: UNAIDS, 2006).

5. UNAIDS, *MDG6: Six Things You Need to Know about the AIDS Response Today* (Geneva: UNAIDS, 2010).

6. McNeil, D. G. "At Front Lines, AIDS War Is Falling Apart," *The New York Times,* 10 May 2010. Accessed 16 June 2010 at www.nytimes.com/2010/05/10/world/africa/10aids.html.

7. UNAIDS and WHO, *AIDS Epidemic Update* (Geneva: UNAIDS and WHO, 2009).

8. UNDP, *Human Development Report* (New York: United Nations Development Programme, 2005).

9. UNAIDS, *AIDS Info: UNAIDS Reference Report* (Geneva: 2010).

10. Commission on AIDS in Asia, *Redefining AIDS in Asia: Crafting an Effective Response* (New Delhi: Oxford University Press, 2008).

11. UNAIDS, *MDG6: Six Things You Need to Know about the AIDS Response Today* (Geneva: UNAIDS, 2010).

12. UNAIDS and WHO. 2009. *Op cit.*

13. National AIDS Control Organisation, *HIV Sentinel Surveillance and HIV Estimation in India 2007: A Technical Brief*, 2008. Accessed 22 June 2010 at www.nacoonline.org/Quick_Links/Publication/.

14. UNAIDS and WHO. 2009. *Op cit.*

15. UNAIDS, *Report on the Global AIDS Epidemic* (Geneva: UNAIDS, 2008).

16. UNAIDS and WHO. 2009. *Op cit.*

17. *Ibid.*

18. Khobotlo, M., R. Tshehlo, J. Nkonyana, M. Ramoseme, et. al., *Lesotho: HIV Prevention Response and Modes of Transmission Analysis* (Maseru, Lesotho: Lesotho National AIDS Commission, 2009); Gelmon, L., P. Kenya, F. Oguya, B. Cheluget, and G. Haile, *Kenya: HIV Prevention Response and Modes of Transmission Analysis* (Nairobi, Kenya: National AIDS Control Council, 2009); Mngadi, S., N. Fraser, H. Mkhatshwa, T. Lapidos, et. al., *Swaziland: HIV Prevention Response and Modes of Transmission Analysis* (Mbabane, Swaziland: National Emergency Response Council on HIV/AIDS, 2009); Bosu, W., K. Yeboah, G. Rangalyan, K. Atuahene, et. al., *Modes of HIV Transmission in West Africa: Analysis of the Distribution of New HIV Infections in Ghana and*

*Recommendations for Prevention* (Accra: Ghana AIDS Commission, 2009); Asiimwe, A., A. Koleros, and J. Chapman, *Understanding the Dynamics of the HIV Epidemic in Rwanda: Modeling the Expected Distribution of New HIV Infections by Exposure Group* (Kigali, Rwanda: National AIDS Control Commission, MEASURE Evaluation, 2009).

19. Wallace, R., "A Synergism of Plagues: 'Planned Shrinkage,' Contagious Housing Destruction, and AIDS in the Bronx," *Environment Research* 47, no. 1 (1988): 1–33.

20. aids2031 Hyper-Endemic Working Group, *Turning Off the Tap: Understanding and Overcoming the HIV Epidemic in Southern Africa* (Johannesburg, South Africa: Nelson Mandela Foundation, 2010. http://www.nelsonmandela.org/index.php/publications/full/turning_off_the_tap_understanding_and_overcoming_the_hiv_epidemic_in_southe/ (Accessed August 6, 2010)

21. Shisana, O., T. Rehle, and L. Simbayi, *South African National HIV Prevalence, HIV Incidence, Behaviour, and Communication Survey* (Cape Town, South Africa: HSRC Press, 2005).

22. UNAIDS, *AIDS Info: UNAIDS Reference Report* (Geneva: 2010).

23. Whiteside, Alan, with Alison Hickey, Jane Tomlinson, and Nkosinathi Ngcobo, "What is driving the HIV/AIDS epidemic in Swaziland? And what more can we do about it?" Report prepared for the National Emergency Response Committee on HIV/AIDS and UNAIDS, Mbabane, April 2003. NERCHA, UNAIDS, and the World Bank Global HIV/AIDS Programme, *Swaziland HIV Prevention Response and Modes of Transmission Analysis*, NERCHA Mbabane Swaziland, March 2009.

24. National Emergency Council on HIV/AIDS, *Swaziland HIV Prevention Response and Modes of Transmission Analysis Study*, National Emergency Response Council on HIV/AIDS (NERCHA) Mbabane Swaziland, March 2009, p. 38.

25. Andrews, Penelope E., *Who's Afraid of Polygamy? Exploring the Boundaries of Family, Equality, and Custom in South Africa*, unpublished paper.

26. NERCHA 2009. *Op cit* p. 40.

27. Whiteside, Alan, and Greg Wood, *Socio-Economic Impact of HIV/AIDS in Swaziland*, Ministry of Economic Planning and Development, Government of Swaziland, 1994.

28. Whiteside, A., and A. Whalley, *Reviewing 'Emergencies' for Swaziland: Shifting the Paradigm in a New Era*, Durban: HEARD, 2007.

29. WHO, UNICEF, and UNAIDS, *Towards Universal Access: Scaling Up Priority HIV/AIDS Interventions in the Health Sector. Progress Report* (Geneva: World Health Organization, 2009).

30. Government of Malawi, *Malawi HIV and AIDS Monitoring and Evaluation Report: 2008–2009*, 2010. Accessed 14 June 2010 at http://data.unaids.org/pub/Report/2010/malawi_2010_country_progress_report_en.pdf.

31. Gelmon, L., P. Kenya, F. Oguya, B. Cheluget, and G. Haile. *Kenya: HIV Prevention Response and Modes of Transmission Analysis* (Nairobi, Kenya: National AIDS Control Council, 2009).

32. Lowndes, C. M., M. Alary, M. Belleau, W. K. Bosu, et. al., *West Africa HIV/AIDS Epidemiology and Response Synthesis: Implications for Prevention* (Washington, DC: World Bank Global HIV/AIDS Program, 2008).

33. UNAIDS and WHO. 2009. *Op cit.*

34. Anderson, R. M., S. Gupta, and W. Ng, "The Significance of Sexual Partner Contact Networks for the Transmission Dynamics of HIV," *Journal of Acquired Immune Deficiency Syndromes* 3, no. 4 (1990): 417–429.

35. Grassly, N. C., C. Fraser, and G. P. Garnett, "Host Immunity and Synchronized Epidemics of Syphilis Across the United States," *Nature* 433 (2005): 417–421.

36. Hall, H. I., R. Song, P. Rhodes, J. Prejean, et. al., "Estimation of HIV Incidence in the United States," *Journal of the American Medical Association* 300, no. 5 (2008): 520–529.

37. Mann, J., and D. J. M. Tarantola, eds., *AIDS in the World II: Global Dimensions, Social Roots, and Responses*, vol. 2 (New York: Oxford University Press, 1996).

38. UNAIDS and WHO. 2009. *Op cit.*

**39.** aids2031 Working Group on Countries in Rapid Transition in Asia, *Asian Economies in Rapid Transition: HIV Now and Through 2031* (Bangkok, Thailand: UNAIDS Regional Support Team for Asia and the Pacific, 2009).

**40.** Statement by Dr HAO Linna, Director-General for International Cooperation of the National Population and Family Planning Commission of China at the General Debates of the 41st Session of the UN Commission on Population and Development (New York: 8 April 2008). Accessed on 30 September 2010 at http://www.un.org/esa/population/cpd/cpd2008/comm2008.htm (China statement on Agenda Item 3).

**41.** aids2031 Working Group on Countries in Rapid Transition in Asia, *Asian Economies in Rapid Transition: HIV Now and Through 2031* (Bangkok, Thailand: UNAIDS Regional Support Team for Asia and the Pacific, 2009).

**42.** UNAIDS and WHO. 2009. *Op cit.*

**43.** Levin, A. M., D. T. Scadden, J. A. Zaia, and A. Krishnan, "Hematologic Aspects of HIV/AIDS," *Hematology* (2001): 463–478.

**44.** Fraser, C., T. D. Hollingsworth, R. Chapman, F. de Wolf, and W. P. Hanage, "Variation in HIV-1 Set-point Viral Load: Epidemiological Analysis and an Evolutionary Hypothesis," *Proceedings of the National Academy of Sciences* 104, no. 44 (2007): 17,441–17,446.

**45.** Cranston, P., and T. Davies, *Future Connect: A Review of Social Networking Today, Tomorrow, and Beyond and the Challenges for AIDS Communicators* (South Orange, N.J.: Communication for Social Change Consortium for aids2031 Communication Working Group, 2009).

**46.** Quinn, D. P., "Capital Account Liberalization and Financial Globalization: A Synoptic View," *International Journal of Finance & Economics* 8 (2003): 189–204.

**47.** Intergovernmental Panel on Climate Change, *Climate Change 2007 Report* (Geneva: IPCC, 2007).

**48.** UNFCC, *Investment and Financial Flows to Address Climate Change* (Bonn, Germany: Climate Change Secretariat, 2007).

**49.** Parry, M., N. Arnell, P. Berry, D. Dodman, et. al., *Assessing the Costs of Adaptation to Climate Change: A Review of the UNFCC and Other Recent Estimates* (London: International Institute for Environment and Development and the Grantham Institute for Climate Change, Imperial College, 2009).

**50.** UNAIDS and WHO. 2009. *Op cit.*

**51.** *Ibid.*

**52.** UNAIDS and WHO. 2008. *Op cit.*

**53.** Izazola-Licea, J. A., J. Wiegelmann, C. Arán, T. Guthrie, P. De Lay, and C. Avila-Figueroa, "Financing the Response to HIV in Low-Income and Middle-Income Countries," *Journal of Acquired Immune Deficiency Syndromes* 52, no. S2 (2009): S119–S126.

**54.** Henry J. Kaiser Family Foundation, *U.S. Federal Funding for HIV/AIDS: The President's Fiscal Year 2011 Budget Request,* 2010. Accessed 25 May 2010 at www.kff.org/hivaids/upload/7029-06.pdf.

**55.** Garnett, G. P., K. Case, T. B. Hallett, J. Stover, and P. K. Piot, "Modeling the HIV Pandemic up to 2031: From an Epidemic to an Endemic Disease," unpublished paper.

**56.** UNAIDS and WHO. 2009. *Op cit.*

**57.** UNAIDS, *AIDS Info: UNAIDS Reference Report* (Geneva: 2010).

**58.** Nigeria National Agency for the Control of AIDS, *United Nations General Assembly Special Session Country Progress Report, Reporting Period January 2008–December 2009.*

# 2

## Generating knowledge for the future

Optimizing long-term results requires generating new knowledge about all aspects of the epidemic. New technologies will be needed to prevent new infections and improve care for those who are infected, and new information is critical to tackling the underlying drivers of the epidemic. Equally important, more efficient management and utilization of knowledge for both policy and programmatic implementation are vital to a sustainable approach.

This chapter examines strategies to generate the breadth of knowledge that will be needed between now and 2031. AIDS illustrates, perhaps better than any other global issue of our time, the potential of research to transform the health and well-being of individuals, families, communities, and entire societies. Yet despite the billions of dollars poured into AIDS research and the numerous resulting advances, many additional breakthroughs are needed to sharply reduce new infections and AIDS deaths. We need to relinquish the faith in an imminent fix that would rapidly cause AIDS to disappear. The world is probably a generation or more away from a workable vaccine, and no cure appears to be available anytime soon.

Despite the sober outlook, additional breakthroughs *are* achievable. Some of these possible breakthroughs have the potential to revolutionize our ability to bring the epidemic under control. This chapter examines some of these potential advances and considers what needs to change to speed their emergence.

## AIDS and the power of science

Scientific breakthroughs have characterized the history of AIDS. Remarkably soon after the appearance of a strange new disease in the early 1980s, investigators definitively characterized the limited means by which the disease could be transmitted.[1] The discovery of the human immunodeficiency virus in 1983 was quickly followed by the development of a blood test to detect it.[2] Researchers also documented some early successes in the ability of focused prevention programs to generate radical changes in sexual behavior, first in gay communities in high-income countries and then among high-risk groups in Brazil and Thailand and among heterosexuals in Uganda.[3]

However, early optimism about these findings quickly gave way to despair, as the epidemic exploded throughout sub-Saharan Africa and large-scale studies demonstrated that the early generation of antiretroviral therapies had failed to extend life.[4] Renewed hope arose with the discovery of the need for combination therapy, and the emergence in the mid-1990s of a new class of antiretroviral drugs—protease inhibitors—which ushered in the era of highly-active antiretroviral therapy (HAART), dramatically extending lives by curtailing viral replication. The development of antiretroviral therapy represents one of the most important medical breakthroughs of the last 50 years. Today, 33 medications from six different antiretroviral classes (including some that combine multiple approved medications) have been brought to market by the pharmaceutical industry.[5]

Initially, it was assumed these treatment breakthroughs would almost exclusively benefit high-income countries. Not only was it expected that the high cost of antiretrovirals would make them unaffordable in low-income countries, but many also argued that it was not feasible to manage a complex chronic disease such as AIDS in countries with poorly functioning health systems. The resistance against introducing antiretroviral therapy in developing countries came from a powerful coalition of development agencies and public health experts. The combined actions of the activist movement, the UN Secretary-General, and UNAIDS, as well as competitive pressure from generic manufacturers, reduced drug costs dramatically, rendering cost-prohibitive medicines much more affordable. The massive use of antiretroviral treatment in low-income countries is an unprecedented example of the introduction of a complex new patented technology supported largely by public funding from high-income countries. By 2010, more than five million people were being treated, a figure that would have been unimaginable only a few years ago. By the end of 2008, HAART had averted 1.4 million deaths in sub-Saharan Africa alone.[6]

Efforts to prevent new infections have also benefited from publicly funded research. A major breakthrough was the demonstration in 1999 that administering a short course of antiretrovirals to pregnant women and their newborns sharply lowered the risk that the infant would become HIV-positive.[7] These findings prompted donors and countries to collaborate in establishing and expanding HIV-prevention programs in antenatal settings; as of December 2008, those programs had averted at least 200,000 new infections among newborns.[8] More recently, three studies in Africa found that male circumcision lowered the risk of men becoming infected during heterosexual intercourse by 60%,[9] leading 13 countries in Africa to begin launching programs to encourage men to be circumcised.[10]

## The limits of science to date

Although such research endeavors have saved millions of lives, scientific tools alone will not ensure a favorable future for the pandemic. Proving that condoms are effective in preventing HIV transmission has not automatically ensured that people will use them. And preventing HIV-positive people from dying may have resulted in an increase in some new infections by rendering the disease less serious in the public mind and inadvertently encouraging an increase in sexual risk-taking.[11]

Researchers have sometimes declared victory prematurely. For example, research findings demonstrating the efficacy of a feasible drug regimen to prevent mother-to-child HIV transmission in low-income settings were rightly hailed as a historic breakthrough. However, in the immediate aftermath of the study's release, significantly less attention focused on the barriers to implementation of this approach in countries with weak antenatal health systems. A decade after the study results were released, most HIV-positive women worldwide still lacked access to regimens to prevent transmission to a newborn child, although in recent years substantial strides have been made in closing the access gap.[12]

A similar approach has been evident with respect to antiretroviral treatments. Existing regimens are a lifelong commitment, and they work well if used with high levels of adherence. Yet a remarkably meager body of research has focused on strategies to maximize adherence to these lifelong regimens, especially in resource-limited settings. Even in high-income countries, estimates suggest that a substantial proportion of people living with HIV do not have access to these medications.[13]

The scientific approach to AIDS demonstrates an unwavering faith in biomedical solutions but also sometimes displays a remarkable blindness about investigating human behavior or the underlying social and cultural factors that affect behavior. Research into structural interventions to reduce vulnerability to HIV has remained almost non-existent. Even as the effects of stigma, social marginalization, gender inequality, and other factors have frustrated AIDS programmers, research has continued to be almost exclusively aimed at biomedicine.

Especially noteworthy is the focus on controlled studies to demonstrate the *efficacy* of new technologies and public health strategies, to the neglect of research on *effectiveness* to inform the application of research advances in the field. As a result, early optimism about research advances has almost invariably been followed by frustration at the slow pace of programmatic implementation. Furthermore, inadequate investments in evaluation studies have often limited the ability of program implementers to integrate new learning to improve their efforts over time.

The remainder of this chapter addresses how generating knowledge for AIDS needs to change to maximize impact over the long term.

## Generating global public goods

The Holy Grail of AIDS research would be a vaccine to prevent infection and a cure for the disease. Such products would represent historic advances, and research efforts to pursue their development merit high priority. In the meantime, however, other new tools are urgently needed to prevent HIV transmission and treat those living with HIV.

When the concept of public goods is mentioned, tangible products such as medicines, diagnostic devices, and other health commodities usually come to mind. But knowledge itself is an essential public good.[14] Making further progress in reducing both deaths and new infections over the next generation depends as much on our knowledge of how to most effectively use the tools we have as on the development of new tools.

Moving forward, substantially greater emphasis should focus on translational, or operational, research that will enable available technologies and strategies to achieve maximum impact.

## Prospects for a cure

If a cure for AIDS were found, it would overcome the problems of lifelong adherence to antiretroviral treatment. It would also restore people living with HIV to full health, a critical need in light of emerging evidence that HIV-associated harm continues to occur even among patients who have achieved the best possible response to treatment. And a cure would interrupt transmission.

One of the key recommendations of the aids2031 Science and Technology Working Group is to "develop better products to treat HIV, with the eventual aim of finding a cure."[15] While an ideal cure would completely clear the body of all traces of the virus, most experts question whether complete eradication in an infected individual is possible, given the ability of the virus to hide in organs and tissues that have remained beyond the reach of existing therapies.[16] Yet it may be possible to find a "functional cure"—that is, one that achieves durable control of the virus in the absence of ongoing antiretroviral therapy.

### New diagnostic technologies

Until a cure is available, the world needs to ensure that antiretroviral therapy is as effective as possible. In high-income countries, HIV clinical practice is characterized by a proliferating array of diagnostic tools that gauge key virologic and immunologic markers. Resistance tests help physicians know which antiretroviral regimens are likely to be most effective for individual patients, and routine monitoring of each patient's viral load and immune function enables clinicians to know when to change drugs to sustain viral suppression and to avert a preventable deterioration of the body's immune system.

**Table 2.1   Point-of-care and low-cost diagnostic development needs in different patient groups[17]**

| Asymptomatic | Symptomatic | On Treatment | Treatment Failure |
|---|---|---|---|
| CD4 test for staging. Test to detect latent TB. Algorithm for diagnosis of pediatric HIV infection using a combination of tests. | CD4 test for staging. Fast diagnosis of pulmonary or extra-pulmonary TB. Fast TB drug sensitivity testing. Lab on a chip to distinguish causative agents of key symptom presentations. | Viral load test that detects viral rebound above the agreed threshold that triggers switching. CD4 count that detects failure to respond immunologically to therapy. Tests to monitor for key antiretroviral toxicities (creatinine, liver enzymes, hemoglobin). | Resistance test that detects specific mutations relevant to second-line drug choice, to inform second-line choice. This option is relevant only if second-line therapy includes drugs from a class already used in first-line therapy. |

Most of these cutting-edge diagnostic tools are unavailable in most low-income settings.[18] The sophisticated equipment needed for diagnosis, such as tests for CD4 counts and resistance to drugs, are designed for well-equipped laboratories and highly trained staff. They are not suitable for the field conditions in which so many people with HIV receive treatment in the developing world. While the U.S. government and other donors have worked to expand access to these diagnostic tools in low-income countries, it is unlikely to be feasible to provide all these technologies to the countless thousands of clinical settings on which people living with HIV depend.

New, simpler, more affordable diagnostic technologies are needed, such as simple tests for use where a patient is being treated that can be delivered on a large scale, at low cost, and used by primary health-care workers with minimal training. Several new point-of-care CD4 counting technologies are currently being evaluated, including the rapid, instrument-free CD4 count from the CD4 Initiative and the PIMA CD4 instrument from Inverness Medical. These tests are low-cost and designed to be used in peripheral health centers with limited laboratory services. Field trial results are expected shortly.

Point-of-care viral load tests present a much more difficult technological challenge and are some years away from introduction, as are point-of-care tests for TB diagnosis and drug sensitivity. However, the development of these tests should be encouraged as simple-to-use point-of-care tools would greatly aid medical decision-making. Even qualitative tools that could identify whether a patient's viral load is suppressed would represent a major step forward in the treatment toolkit for HIV. Such a qualitative test would have an enormous impact on early infant diagnosis. Recent experience with rapid HIV testing technologies indicates that developing countries will put appropriately designed diagnostic tools to use if and when they are made available.

Recent studies demonstrate that excellent clinical results are achievable even when state-of-the-art diagnostic tools are not available, especially if there is intensive clinical monitoring.[19] However, robust, affordable diagnostic assays remain an urgent research priority, to enable clinicians to determine when to switch regimens and to minimize the transmission of drug-resistant virus.

Another key recommendation of the aids2031 Science and Technology Working Group is to launch a global initiative to coordinate the development of diagnostic and monitoring tools for HIV, which they feel is an urgent need to bridge what they call the "diagnostic divide."[20]

### New prevention tools

Whereas greatly expanded coverage of current HIV prevention interventions may further reduce HIV incidence by 50% (see discussion in Chapter 1, "The future of AIDS: a still-unfolding global challenge"), new tools will be necessary to virtually halt the spread of HIV in the most affected populations such as in Southern Africa or in concentrated populations such as among men who have sex with men. Such tools would include a microbicide, pre-exposure prophylaxis (PrEP), a vaccine, a cure, and possibly high coverage with early antiretroviral therapy.

In the quest to develop new prevention tools, researchers have identified several potentially promising approaches that involve the use of antiretrovirals to prevent transmission:[21]

- **Pre-exposure prophylaxis (PrEP) and microbicides**— Evidence from animal studies suggests that it may be possible to reduce the risk of HIV transmission by taking antiretrovirals before a possible exposure to the virus. Eight human trials are currently underway to study the pre-exposure use of antiretrovirals to reduce the risk of HIV transmission. Microbicides are compounds that can be formulated in gels, films, rings, or sponges. Most microbicide candidates are intended for vaginal application, although some are also being studied to protect against transmission during anal intercourse.

Following several failed attempts with other microbicides, investigators from Durban, South Africa, demonstrated in the so-called CAPRISA trial that a vaginal gel with 1% tenofovir (an antiretroviral drug used for treating people with HIV) reduces the risk for women of acquiring HIV during sex by 39%, and about 54% in those who used the gel consistently.[22] This study provides a first proof-of-concept for two prevention approaches: the use of antiretrovirals to prevent sexual transmission of HIV (as already used for the prevention of mother-to-child transmission) and the use of a topical gel for the same.

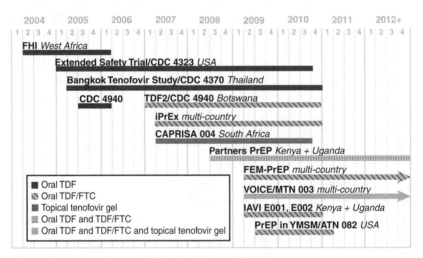

Source: AIDS Vaccine Advocacy Group (AVAC) May 2010

Figure 2.1    Past PrEP trials and estimated end-dates for ongoing trials. Note: Trial end-dates are estimates. Due to the nature of clinical trials, the actual dates may change.

These results need to be confirmed in a large trial and further studies are needed to determine optimal regimens. This will take several years, while other oral PrEP regimens are being evaluated in diverse populations. While highly encouraging, the CAPRISA trial should also remind us of the complexity, cost, and time needed for such trials.

- **Treatment as prevention**—By suppressing the HIV virus in people who are infected, antiretroviral therapy may dramatically reduce infectivity and slow the rate of new infections. A number of studies support this hypothesis, including one that found that HIV transmission in serodiscordant couples—in which one partner is HIV positive and the other is not—was reduced when the infected partner was receiving antiretroviral therapy.[23] In late 2008, a team of WHO scientists released the results of a modeling exercise that suggested that providing annual HIV testing combined with free antiretroviral therapy to all who test HIV positive could sharply lower annual HIV incidence within a decade and lay the foundation for the eventual elimination of HIV.[24] Modeling suggests that, under these assumptions, initiating antiretroviral therapy earlier in the course of infection would significantly lower the number of incident infections, as would significant increases in treatment coverage. Faster scaling-up of treatment would accelerate the assumed prevention gains associated with antiretrovirals. Numerous studies are now underway to test various aspects of this so-called "test and treat" approach, although initiation of large-scale efficacy or effectiveness trials is at least several years away and would entail formidable implementation challenges.

Each of these research avenues holds promise, but well-designed studies are required to support their widespread use. Numerous experimental HIV prevention options have been grounded in strong evidence of plausibility, only to have studies yield disappointing results. Based on evidence of women's physiological vulnerability to HIV, it was once thought that a physical cover for the cervix could reduce the risk of transmission, but studies of female diaphragms did not find the approach efficacious for HIV prevention.[25] Similarly, although epidemiological studies have consistently found a strong association between HIV infection and herpes simplex virus type 2 (HSV-2 ),[26] evaluations of acyclovir suppressive therapy for HSV-2 have failed to demonstrate an HIV prevention benefit.[27] Regardless of the eventual outcome for the current leading candidates for HIV prevention, it is imperative that researchers continue to innovate and push the field forward.

In doing so, researchers should ensure that trials have sufficient power to obtain answers regarding the efficacy of experimental prevention methods. Many trials of experimental prevention technologies have failed to yield definitive results.

Proof of efficacy is, of course, an essential step in the development of new prevention tools. However, additional knowledge will be needed to guide the use of new tools. For new prevention technologies, which will inevitably confer only partial protection, new social norms and patterns of behavior may emerge that could increase or decrease new infections. For instance, the risk exists that sexually active people will erroneously view any new tool as a foolproof shield against infection, rendering superfluous the need to use a condom or reduce the number of sex partners. Alternatively, new prevention tools might encourage greater attention to risk reduction. Community education programs will continue to be needed as new tools are rolled out, and longer-term monitoring will be needed to detect unforeseen consequences of new prevention strategies.

### A preventive vaccine

When Margaret Heckler, then the U.S. Secretary of Health and Human Services, reported on the discovery of HIV in 1983, she boldly predicted that a preventive vaccine would be developed within two years. Despite this early optimism, progress towards a vaccine has been slow.

After extensive scientific experimentation, it is widely accepted that the development of an HIV vaccine is one of the most difficult challenges confronting biomedical research. An ideal vaccine would be feasible for widespread use in low-income settings, confer lifelong immunity, protect against all routes of HIV transmission, and work against diverse viral strains and subtypes.

In 2009, a trial in Thailand was the first large human vaccine efficacy trial to show promise, with the candidate, a prime-boost regimen of two genetically engineered vaccine components, demonstrating modest protection against infection.[28] What seems most likely is that future progress will be incremental; more effective products will supersede earlier vaccines over a considerable period of time. The limited efficacy of early HIV vaccines means they would most likely have to be used as complementary tools in combination with existing prevention strategies. Funders and policymakers must understand that several generations of vaccine development are needed before a product emerges that is capable of ending the epidemic. Product development, animal testing, and human efficacy trials are all expensive, but the health of future generations demands robust and sustained investments. Modeling indicates, for instance, that a vaccine with 70% efficacy provided to 40% of the population beginning in 2015 could reduce the annual number of infections by 81% by 2030.[29]

The urgent need for an effective vaccine, combined with the painfully slow progress achieved to date, suggests that standard approaches to biomedical research may need to be supplemented by a more focused and strategic approach. Traditionally, public sector agencies fund basic and applied medical research, typically awarding grants to independent scientific teams that compete among themselves to come forward with the most promising ideas. With the possibility of financial rewards for future patents, intellectual property is carefully guarded, with independent research teams competing to show the world that their approach is superior to all others. This traditional approach has repeatedly paid dividends, generating an array of antiretroviral drugs that have saved the lives of millions. But it has yet to produce tangible progress toward an AIDS vaccine.

The Global HIV Vaccine Enterprise, a multistakeholder partnership uniting key players in the AIDS vaccine field, has worked to enhance collaboration to solve key scientific challenges. Although this approach has facilitated dialogue among researchers in the field, it has not been able to overcome the turf-consciousness of many leading players.

Recent history suggests that other approaches are possible when it comes to mission-driven research undertakings. International collaboration resulted in a remarkably swift mapping of the human genome, and CERN's multicountry collaborative efforts aim to answer key questions of physics. For an enterprise as urgent as the search for a preventive AIDS vaccine, a similar mission-driven approach is needed. Research funding should be provided on condition that all research findings are shared. Efforts should be made to avoid the duplication of work and the lack of communication among different players. This strong commitment to vaccine research should investigate multiple approaches simultaneously, to avoid the risk that the failure of a single vaccination theory could cause the entire research edifice to come crashing down.

## Building the evidence base for community-level and structural prevention approaches

The continuing focus on developing new prevention technologies reflects both recognition of the need for a combination of approaches and some disappointment and skepticism regarding the long-term effectiveness of strategies to change sexual behaviors. Behavior change has been central to every major HIV prevention success ever recorded. But changing behaviors is a difficult, complex, lifelong endeavor. Furthermore, the success of approaches that have proven efficacious in controlled clinical trials has often been difficult to replicate in the real world.

Existing models for behavioral HIV-prevention programs are grounded in a rather narrow spectrum of cognitive behavioral theories. These strategies focus on the individual as the fulcrum for prevention success. Even when these behavioral strategies work, their effectiveness is often short lived. And to the extent that behavioral strategies are effective, their lessons have to be heeded one individual at a time, one sexual episode at a time.

By contrast, when entire societies change and adopt new social norms, these norms tend to be self-perpetuating and self-policing. Social norms are backed by formal and informal social sanctions, and groups develop ways to support one another in adhering to these norms.[30] With a mindset demanding immediate results, the AIDS movement has often adopted a reductionist approach that focuses on individual behaviors, regarding social change as a "luxury" item that requires too much time to achieve impact. As the aids2031 Working Group on Social Drivers concluded, a long-term focus on sustainability recognizes social forces as *fundamental* to success against the epidemic.[31] As part of this approach, it is more useful to think in terms of "practices" than "behaviors," as the former conveys the social dimensions of unprotected intercourse, sexual concurrency, sharing of syringes, or human reproduction.[32]

Notwithstanding the centrality of social forces to the continued perpetuation of the pandemic, remarkably little research has focused on strategies to marshal social forces to prevent new infections. The overwhelming majority of behavioral intervention studies have evaluated individual-level or small group interventions, with scant attention paid to societal-level interventions.

Nor is the evidence base well-developed for structural or public policy interventions to reduce vulnerability.[33] For example, the effects of gender inequality on women's risk and vulnerability are plain in many countries, yet, to date, only *one* structural intervention that combines women's empowerment with HIV prevention has been rigorously evaluated to assess its impact on the rate of new HIV infections.[34] Similarly, although participatory research methods have yielded overwhelming consensus that HIV stigma is a primary impediment to an effective AIDS response, the evidence base for antistigma programming remains weak.

For every strategy that seeks to promote HIV prevention, reduced HIV incidence is the desired endpoint. But structural interventions and social change processes may require a decade or more to generate actual reductions in new infections. Their longer-lasting and self-reinforcing qualities make them critical to a long-term approach to AIDS, but building the evidence base for action demands knowledge-generation approaches that differ from the time-limited clinical trials on which the AIDS field has traditionally focused.

Articulation and testing of the anticipated causal pathway for individual epidemic drivers will help guide research on key social factors. By generating evidence relevant to particular settings at key steps along a causal pathway, it is possible to steer programs to better contribute to the ultimate goal: a reduction in new HIV infections.

### Innovative financing and mission-oriented research and development

Strong, sustained funding from the public sector is needed to help generate the knowledge required to develop new tools and technologies. An analysis commissioned by the Bill & Melinda Gates Foundation concluded that, in 2007, global research and development funding related to AIDS in resource-limited settings amounted to US$ 1.083 billion (of which 63% was devoted to vaccines, 18% to microbicides, and 1.2% to diagnostics). In the United States, the research funding has flattened, and Europe has an even greater lack of public funding. Clearly, these trends must be reversed if scientists are to have the resources they need to generate new approaches to fighting AIDS.

Whereas the public sector specializes in supporting research to generate basic science advances, product development expertise generally lies with private industry. As long as a robust market for antiretroviral drugs exists in high-income countries, the private sector has a strong financial incentive to invest in research and development on new therapeutics. However, industry's incentive is less clear when it comes to simple, low-cost tools for primary use in resource-limited settings.

When market dynamics do not align with international needs, innovative approaches are required. Two common approaches to the problem involve the use of so-called "push" and "pull" mechanisms. Push mechanisms provide a direct nudge to private companies, offering them up-front financial support to pursue promising research avenues. Pull mechanisms seek to affect the private sector's calculus regarding likely future profits associated with new products or health tools, effectively "pulling" them to invest in research approaches that might not otherwise seem promising.

AIDS has helped give rise to a common push mechanism: the public-private product development partnership that seeks to combine the public sector's commitment to international public goods with the private sector's business approach to product development. Public-private partnerships in the AIDS field include the International AIDS Vaccine Initiative (IAVI) and the International Partnership for Microbicides (IPM). IAVI and IPM provide direct grants or enter into license agreements with private companies to encourage their engagement in AIDS research. Although it is too soon to know whether either IAVI or IPM has actually accelerated progress toward their respective product goals, both have mobilized new resources for AIDS research, increased awareness of the need for new health tools, and engaged pharmaceutical companies that were previously uninvolved in AIDS work.

An important question is whether a similar public-private partnership will be needed in the future for antiretroviral products. To date, little evidence of market failure for antiretrovirals exists. However, even with dramatic declines in the prices of antiretrovirals, these drugs remain far more expensive than therapeutic agents normally used in resource-limited settings. Sustaining treatment access in developing countries will require new, less expensive approaches to HIV treatment, such as the development of easy-to-administer compounds that need to be taken only periodically rather than daily.

## Innovating to spur investments in prevention research for women

As HIV infections among women steadily increased during the 1990s, it became increasingly apparent that the lack of prevention options for women was compounding their vulnerability to HIV. Experts had long recognized that vaginal microbicides that women themselves could control had enormous potential to stem the epidemic's spread. Yet efforts to develop one or more safe and effective microbicides for the prevention of HIV transmission attracted only low-level support during the 1990s from public-sector research agencies and commercial entities.

Leading foundations and global health experts joined together in 2002 to establish the International Partnership for Microbicides. Similar to IAVI and other product-oriented, public–private partnerships in the health field, IPM manages funds from public and philanthropic sectors to spur development of health products needed in developing countries. As with similar organizations, IPM is a lean operation, using only 13% of its funding for administration.

In its first seven years of existence, IPM obtained six nonexclusive licenses from major pharmaceutical companies to test potentially promising compounds as microbicides. As of December 2009, IPM had completed or initiated more than 25 scientific studies. To support current and future research on microbicides, IPM has established a global research network of more than 20 sites in seven African countries. IPM alone accounts for more than 20% of all noncommercial funding for microbicide research and development.

Pull mechanisms seek to create confidence that a profitable market for a new product will exist. One innovative pull mechanism is the *advance market commitment*, whereby donors make an advance commitment to buy a certain quantity of a product at an agreed price once the product is available for use. The first advance market commitment focused on a new pneumococcal vaccine. Experience with the pneumococcal vaccine will indicate whether the advance market commitment approach might work for other tools, such as AIDS vaccines, microbicides, or point-of-care diagnostics.

Innovative funding mechanisms, such as the Global Fund to Fight AIDS, Tuberculosis and Malaria, and the Global Alliance for Vaccines and Immunization, also function as "pull" mechanisms, in that their existence demonstrates global resolve to bring international public goods to those who need them. The potential impact of these mechanisms on industry's thinking about investment decisions is yet another important reason to ensure their survival and success.

Regardless of the method of financing used to spur greater research investments, it is apparent that the overwhelming reliance on research centers in high-income countries to conduct AIDS studies in developing countries makes little sense for the long term. In reality, a number of emerging economies—including Brazil, China, India, South Africa, and Thailand—have rapidly growing domestic research capacities. International donors, developing country governments, foundation funders of medical research, and international technical agencies should collaborate to establish research centers of excellence in low- and middle-income countries. This approach is in widespread use with agricultural research.

### A new approach to programmatic research

The AIDS response has generated a remarkable array of validated prevention and treatment strategies, yet almost invariably, results in the real world have fallen short of expectations based on the results of controlled clinical trials. Virtual elimination of mother-to-child HIV transmission may indeed be feasible with existing tools, yet in 2009, an estimated 370,000 children became infected.[35] Based on efficacy studies, modeling suggests that available strategies could prevent two out of three new HIV infections—but in 2009, 2.6 million people became newly infected with HIV.[36] Similarly, even though antiretroviral therapies have the capacity to prevent or delay death for the vast majority of people who receive them, 1.8 million people died of AIDS in 2009.[37]

Limited access to services is undoubtedly a major impediment to realizing the potential of existing tools, yet it is hardly the only obstacle. The AIDS field has adopted a remarkably narrow approach to research, focusing almost exclusively on the ideal results that may be achieved in controlled settings. Until AIDS research focuses as much energy on the real world as it does on clinical trials, we will inevitably be disappointed with the results.

## From efficacy to effectiveness

Studies have documented the efficacy of a broad range of strategies to encourage individuals to adopt safer sexual behaviors.[38] But these effects have been difficult to replicate in community settings. In part, this reflects the difference between the carefully controlled environment of a clinical trial and the real world. In clinical trials, participants are selected based on rigorous criteria, but actual programs are typically offered to a much broader range of individuals. With an eye toward isolating the specific benefit of a particular programmatic intervention, clinical studies tend to study single-dimension interventions, yet actual HIV-prevention programs typically combine multiple elements, such as counseling, social marketing, HIV testing promotion, and sexually transmitted infection (STI) screening and treatment.

Efficacy studies are also time-limited. Few HIV-prevention studies follow participants longer than 12 months, with many basing their conclusions on effects reported immediately after receipt of the intervention. One study found that a 10-week individualized counseling program for men who have sex with men resulted in significant positive changes in self-reported sexual behaviors and short-term reductions in HIV incidence. Concerned that most prevention trials followed study participants only for short periods, study investigators followed up more than three years later to assess the results. What they found was startling. The favorable behavior changes detected soon after the end of the 10-week program appeared to dissipate 12 to 18 months later. More than three years after the program ended, there was no difference in HIV infection rates between recipients of the intervention and the study control group that had not received the intervention.[39]

With few exceptions, behavioral intervention trials rely on participants' self-reported knowledge, behaviors, and future intentions.[40] Although methods have been developed to increase the reliability of self-reported behaviors, self-reports are inevitably susceptible to biases and poor recall.

Observational studies in different settings have identified declines in HIV incidence that exceed what would be expected from natural saturation of infection,[41] suggesting that programmatic approaches may be having an effect. But the lack of well-planned effectiveness studies makes it impossible to draw definitive conclusions or to attribute trends to particular approaches. Substantially greater attention must be directed toward rigorous effectiveness studies.

For both prevention and treatment programs, greater investments are needed in operational or translational research to guide program implementers. Chapter 3, "Using knowledge for a better future," addresses this priority in greater detail. Another related topic that Chapter 3 addresses is the need for comprehensive monitoring of program inputs, outputs, and costs to promote efficient and effective service delivery, and the need to evaluate the impact of interventions and combinations of interventions to measure (and improve) their effectiveness.

## Measuring new HIV infections

Building the evidence base to monitor the impact of HIV-prevention efforts confronts a considerable obstacle. Presently, no reliable test exists to measure new HIV infections. Several techniques have been developed to differentiate recent HIV infections from longstanding infections,[42] but use of these methods in various low- and middle-income countries has identified various weaknesses, such as misclassification of some individuals who had been infected for a long time and others who were receiving antiretroviral therapy.[43]

### Generating knowledge to sustain HIV treatment

Not all antiretroviral regimens are created equal. According to national treatment monitoring in the United Kingdom, the duration of treatment success varies up to threefold, depending on which first-line regimen is administered.[44] It is important to select the longest-lasting first-line regimen, especially for low-income countries where second- and third-line regimens are likely to be unaffordable and unavailable. Effective first-line regimens delay or avert the much-higher costs associated with second-line drugs and, in settings where second-line therapy is not available, they can keep patients alive and healthy until future research breakthroughs can deliver more affordable options.

Unfortunately, clear evidence does not exist regarding the optimal, longest-lasting regimen suitable for use in resource-limited settings. This is a critical research priority to enable program planners in low-income countries to use limited resources to maximize individual health and longevity.

Another area where better knowledge is needed centers on simplifying and decentralizing the administration of antiretroviral treatment to the greatest extent that is consistent with sound clinical outcomes. In a recent randomized trial in Uganda, researchers found that home-based care delivered by community workers was at least as effective as clinic-based programs.[45]

Sustaining treatment gains also requires building the knowledge base for promoting strong treatment adherence. The relatively few studies that have assessed HIV treatment adherence have followed patients for short periods of time. A recent, longer study in London found that patients in the U.K. are often able to maintain high adherence levels a decade or more after initiating treatment.[46] Additional studies that use robust research methods are needed to build the evidence base for interventions to support treatment adherence in low- and middle-income countries.

### Strengthening local knowledge

AIDS epidemics can vary considerably from country to country, as well as within countries, underscoring the need for locally specific and relevant data.

Timely and accurate information about the epidemic is the starting point for sound planning of HIV strategies and programs. Knowing who is becoming infected and at what rate, as well as what practices and factors are driving the epidemic, is important. To plan treatment programs, it is important to understand where and at what rate HIV infections are occurring, the size and distribution of gaps in testing services, the number of people who immediately need therapy, the likely number of people who will require treatment in future years, the relative need for services among populations or geographic districts, the success of different clinical settings in retaining patients in care, the rate of treatment failure, and details about the emergence of viral strains that are resistant to standard antiretroviral regimens.

Researchers have developed a range of strategies to use limited data from public health surveillance to estimate new HIV infections in different settings.[47] One of the most promising is the "estimating HIV incidence by modes of transmission" model, which uses available data to develop a single-year estimate of the number and distribution of new

HIV infections in a given country.[48] In 2008–2009, with support from UNAIDS, epidemiological syntheses by modes of transmission were conducted in 26 countries.[49] UNAIDS plans to extend this approach to at least 30 additional countries in 2010–2011.

These recent epidemiological assessments have provided critical insights to inform programmatic priorities. A comparison of epidemiological trends in Ghana and Swaziland is instructive (see Figure 2.2). Both are African countries, yet the factors driving their respective epidemics are notably different. HIV transmission during sex work accounts for a considerably larger share of new infections in Ghana than in Swaziland, where the proportion of incident infections among individuals in stable heterosexual relationships is substantially greater. Clearly, Ghana and Swaziland need to pursue national AIDS responses that differ markedly from one another if they are to address their respective national priorities and maximize long-term impact.

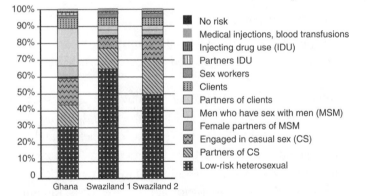

Source: WHO/UNAIDS AIDS epidemic update 2009.

Figure 2.2   Distribution of new infections by mode of exposure in Ghana and Swaziland, 2008. (Note: Sensitivity analysis for Swaziland used different data sources.)

Periodic assessments of new infections by modes of transmission must become standard operating procedure for AIDS planning. Additional efforts are needed to improve and standardize the methodology for such reviews, to allow for comparisons across different geographic locations and over time.

Epidemiological assessments are essential to, but not sufficient in, informing national and subnational decision-making on AIDS. The AIDS response has badly neglected social science assessments. If countries are to address the drivers of their epidemics, they need to supplement assessments of the modes of transmission for incident infections with routine sociological assessments, to identify and explore the dimensions of social context that increase risk and vulnerability.[50, 51]

The lack of country-specific sociological assessments often leads to assumptions about the drivers of national epidemics that are not supported by evidence. For example, it is frequently said that poverty is a major cause of AIDS. A correlation may exist between household poverty and vulnerability to HIV in some settings, such as Burkina Faso, but evidence from numerous countries in sub-Saharan Africa fails to identify a strong association between wealth and HIV prevalence (see Figure 2.3). Likewise, it is frequently asserted as a universal proposition that gender inequality increases women's vulnerability to HIV; although strong evidence for this proposition arises in many African countries, in many settings worldwide, women have few rights but HIV prevalence is significantly higher among men. Only rigorous setting-specific assessments can help ensure that national strategies are based on sound evidence rather than received wisdom.

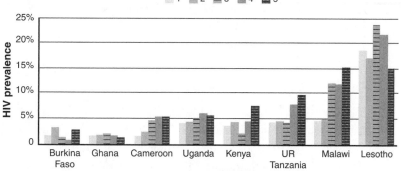

*Source: Report on the Global AIDS Epidemic, UNAIDS 2008.*

Figure 2.3   HIV prevalence in men by wealth status. Quintiles 1–5 represent the poorest 20% of population (1) through to the richest 20% (5), increasing in 20% increments.

An extraordinary body of knowledge already exists on what approaches work best both for prevention of new infections and effective treatment for people who are living with HIV. But, as discussed in this chapter, while the use of existing tools can be improved significantly, new tools are needed to achieve radical reductions in new infections and AIDS deaths by 2031.

The AIDS field has made some of the most important scientific advances of the last 30 years. But only new ways of thinking and operating can generate the knowledge that will be needed to drive innovation and achieve results over the next generation.

Ultimately, knowledge is meaningful only to the degree that it is put to effective use. The next chapter explores how best to translate knowledge into effective, sustainable action.

## Endnotes

1. Curran, J. W., W. M. Morgan, A. M. Hardy, H. W. Jaffe, W. W. Darrow, and W. R. Dowdle, "The Epidemiology of AIDS: Current Status and Future Prospects," *Science* 229, no. 4,720 (1985): 1,352–1,357.

2. Merson, M. H., J. O'Malley, D. Serwadda, and C. Apisuk, "HIV Prevention 1: The History and Challenge of HIV Prevention," *The Lancet* 372, no. 9,637 (2008): 475–488.

3. Global HIV Prevention Working Group, "Behavior Change and HIV Prevention: (Re)considerations for the 21st Century," 2008. Accessed on 16 June 2010 at www.globalhivprevention.org/reports.html.

4. Aboulker, J. P. and A. M. Swart, "Preliminary Analysis of the Concorde Trial," *The Lancet* 341, no. 8,849 (1993): 889–890.

5. U.S. Food and Drug Administration, "Antiretroviral Drugs Used in the Treatment of HIV Infection," 2010. Accessed 10 May 2010 at www.fda.gov/ForConsumers/byAudience/ForPatientAdvocates/ HIVandAIDSActivities/ucm118915.htm. Alcorn, S. *HIV/AIDS Technologies: A Review of Progress to Date and Current Prospects,* working paper, aids2031 Science and Technology Working Group, Seattle, 2008.

6. UNAIDS and WHO, *AIDS Epidemic Update* (Geneva: UNAIDS and WHO, 2009).

7. Guay, L. A., P. Musoke, T. Fleming, D. Bagenda, et al., "Intrapartum and Neonatal Single-Dose Nevirapine Compared with Ziduovudine for Prevention of Mother-to-Child Transmission of HIV-1 in Kampala, Uganda: HIVNET 012 Randomized Trial," *The Lancet* 354, no. 9,181 (1999): 795–802.

 8. UNAIDS and WHO. 2009. *Op cit.*

 9. Gray, R. H., G. Kigozi, D. Serwadda, F. Makumbi, et al., "Male Circumcision for HIV Prevention in Men in Rakai, Uganda: A Randomized Trial," *The Lancet* 369, no. 9,562 (2007): 657–666; Bailey, R. C., S. Moses, C. B. Parker, K. Agot, et al., "Male Circumcision for HIV Prevention in Young Men in Kisumu, Kenya: A Randomized Controlled Trial," *The Lancet* 369, no. 9,562 (2007): 643–656; Auvert, B., D. Taljaard, E. Lagarde, J. Sobngwi-Tambekou, R. Sitta, and A. Puren, "Randomized, Controlled Intervention Trial of Male Circumcision for Reduction of HIV Infection Risk: The ANRS 1265 Trial," *PLoS Medicine* 2, no. 11 (2005): e298.

10. WHO, UNICEF, and UNAIDS, *Towards Universal Access: Scaling Up Priority HIV/AIDS Interventions in the Health Sector,* progress report (Geneva: World Health Organization, 2009).

11. Ostrow, D. E., K. J. Fox, and J. S. Chmiel, "Attitudes Towards Highly Active Antiretroviral Therapy Are Associated with Sexual Risk Taking Among HIV-Infected and Uninfected Homosexual Men," *AIDS* 16, no. 5 (2002): 775–780.

12. WHO, UNICEF, and UNAIDS, *Towards Universal Access: Scaling Up Priority HIV/AIDS Interventions in the Health Sector,* progress report (Geneva: World Health Organization, 2009).

13. Fleming, P., R. H. Byers, P. A. Sweeney, D. Daniels, J. M. Karon, and R. S. Janssen, "HIV Prevalence in the United States," Ninth Conference on Retroviruses and Opportunistic Infections, 2002, Seattle, WA.

14. Stiglitz, J. E., "Knowledge As a Global Public Good," in *Global Public Goods: International Cooperation in the 21st Century,* edited by I. Kaul, I. Grunberg, and M. Stern (New York: Oxford University Press, 1999).

15. Aids2031 Science and Technology Working Group Report. *Advancing science and technology to change the future of the AIDS pandemic.* Seattle: Program for Appropriate Technology (PATH) and Duke University, 2010. http://www.aids2031.org/working-groups/science-and-technology (Accessed August 6, 2010).

16. Deeks, S. G., "HIV Eradication: Is It Feasible?" working paper, aids2031 Science and Technology Working Group, Seattle, Wash., 2008.

17. *Aids2031 Science and Technology Working Group Report: Advancing Science and Technology to Change the Future of the AIDS Pandemic* (Seattle: Program for Appropriate Technology [PATH] and Duke University, 2010).

18. Gerlach, J., D. Boyle, G. Domingo, B. Weigl, and M. Free, "Increased Access to Diagnostic Tests for HIV Case Management," working paper, aids2031 Science and Technology Working Group, Seattle, Wash., 2008.

19. DART Trial Team, "Routine Versus Clinically Driven Laboratory Monitoring of HIV Antiretroviral Therapy in Africa (DART): A Randomized Non-inferiority Trial," *The Lancet* 375, no. 9,709: 123–131.

20. Aids2031 Science and Technology Working Group Report. *Advancing science and technology to change the future of the AIDS pandemic.* Seattle: Program for Appropriate Technology (PATH) and Duke University, 2010. http://www.aids2031.org/working-groups/science-and-technology (Accessed August 6, 2010).

21. Mastro, T. D., W. Cates, and M. S. Cohen, "Antiretroviral Products for HIV Prevention: Looking Toward 2031," working paper, aids2031 Science and Technology Working Group, Seattle, 2008.

22. Karim Q.A., et al, "Effectiveness and Safety of Tenofovir Gel, an Antiretroviral Microbicide, for the Prevention of HIV Infection in Women," *Sciencexpress* 20 (July 2010).

23. Donnell, D., J. M. Baeten, J. Kiarie, K. K. Thomas, et al., "Heterosexual HIV-1 Transmission After Initiation of Antiretroviral Therapy: A Prospective Cohort Analysis," *The Lancet* 375, no. 9,731: 2,092–2,098.

24. Granich, R. M., C. F. Gilks, C. Dye, K. M. De Cock, and B. G. Williams, "Universal Voluntary HIV Testing with Immediate Antiretroviral Therapy As a Strategy for Elimination of HIV Transmission: A Mathematical Model," *The Lancet* 373, no. 9,657: 48–57.

25. Padian, N. S., A. van der Straten, G. Ramjee, T. Chipato, et al., "Diaphragm and Lubricant Gel for Prevention of HIV Acquisition in Southern African Women: A Randomized Controlled Trial," *The Lancet* 370, no. 9,583: 251–261.

26. Corey, L., "Synergistic Copathogens—HIV-1 and HSV-2," *The New England Journal of Medicine* 356, no. 8: 854–856.

27. Celum, C., A. Wald, J. R. Lingappa, A. S. Magaret, et al., "Acyclovir and Transmission of HIV-1 from Persons Infected with HIV-1 and HSV-2," *The New England Journal of Medicine* 362, no. 5 (2010): 427–439.

28. Rerks-Ngarm, S., P. Pitisuttihum, S. Nitayahpan, J. Kaewkungwal, et al., "Vaccination with ALVAC and AIDSVAX to Prevent HIV-1 Infection in Thailand," *The New England Journal of Medicine* 361, no. 23: 2,209–2,220.

29. Stover, J., L. Bollinger, R. Hecht, C. Williams, and E. Roca, "The Impact of an AIDS Vaccine in Developing Countries: A New Model and Initial Results," *Health Affairs* 26, no. 4 (2007): 1,147–1,158.

30. Auerbach, J. D., J. O. Parkhurst, C. F. Cáceres, and K. E. Keller, "Addressing Social Drivers of HIV/AIDS: Conceptual, Methodological, and Evidentiary Considerations," working paper, aids2031 Social Drivers Working Group. Accessed 4 October 2010 at http://www.aids2031.org/pdfs/aids2031%20social%20drivers%20paper%2024-auerbach%20et%20all.pdf.

31. aids2031 Social Drivers Working Group, "Revolutionizing the AIDS Response: Enhancing Individual Resilience and Supporting AIDS Competent Communities," synthesis paper, aids2031 Social Drivers Working Group, Clark University, Worcester, Mass., 2010.

32. Auerbach, et al. *Op cit.*

33. Gupta, G. R., J. O. Parkhurst, J. A. Ogden, P. Aggleton, and A. Mahal, "HIV Prevention 4: Structural Approaches to HIV Prevention," *The Lancet* 372, no. 9,640 (2008): 764–775.

34. Pronyk P. M., J. R. Hargreaves, J. C. Kim, L. A. Morison, et al., "Effect of a Structural Intervention for the Prevention of Intimate-Partner Violence and HIV in Rural South Africa: A Cluster Randomized Trial," *The Lancet* 368, no. 9,551): 1,973–1,983.

35. UNAIDS, *AIDS Info: 2010 UNAIDS Reference Report* (Geneva: 2010).

36. *Ibid.*

37. *Ibid.*

38. Global HIV Prevention Working Group, "Behavior Change and HIV Prevention: (Re)considerations for the 21st Century," 2008. Accessed 16 June 2010 at www.globalhivprevention.org/reports.html.

39. *Ibid.*

40. Lyles, C. M., L. S. Kay, N. Crepaz, J. H. Herbst, et al., "Best-Evidence Interventions: Findings from a Systematic Review of HIV Behavioral Interventions for U.S. Populations at High Risk, 2000–2004," *American Journal of Public Health* 97, no. 1 (2007): 133–143

41. Gregson, S., G. P. Garnett, C. A. Nyamukapa, T. B. Hallett, et al., "HIV Decline Associated with Behavior Change in Eastern Zimbabwe," *Science* 311, no. 5,761 (2006): 664–666; Stoneburner, R. L, and D. Low-Beer, "Sexual Partner Reductions Explain Human Immunodeficiency Virus Declines in Uganda: Comparative Analyses of HIV and Behavioral Data in Uganda, Kenya, Malawi, and Zambia," *International Journal of Epidemiology* 33 (2004): 1–10.

42. Hall, H. I., R. Song, P. Rhodes, J. Prejean, Q. An, et al., "Estimation of HIV Incidence in the United States," *Journal of the American Medical Association* 300, no. 5 (2008): 520–529.

43. Hargrove, J. W., J. H. Humphrey, K. Mutasa, P. H. Parekh, et al., "Improved HIV-1 Incidence Estimates Using the BED Capture Enzyme Immunoassay," *AIDS* 22, no. 4 (2008): 511–518.

44. Beck, E. J., S. Mandalia, M. Youle, R. Brettle, et al., "Treatment Outcome and Cost-Effectiveness of Different Highly Active Antiretroviral Therapy Regimens in the U.K. (1996–2002)," *International Journal of STD & AIDS* 19 (2008): 297–304.

45. Jaffar, S., B. Amuron, S. Foster, J. Birungi, et al., "Rates of Virological Failure in Patients Treated in a Home-Based Versus a Facility-Based HIV-Care Model in Jinja, Southeast Uganda: A

Cluster-Randomized Equivalence Trial," *The Lancet* 374, no. 9,707 (2009): 2,080–2,089.

46. Cambiano, V., F. C. Lampe, A. J. Rodger, C. J. Smith, et al., "Long-Term Trends in Adherence to Antiretroviral Therapy from Start of HAART," *AIDS* 24, no. 8 (2010): 1,153–1,162.

47. UNAIDS and WHO. *Op cit.*

48. Guows, E. P., White, J. Stover, T. Brown, et al., "Short Term Estimates of Adult HIV Incidence by Modes of Transmission: Kenya and Thailand As Examples," *Sexually Transmitted Infections* 82 (2006): iii51–iii55.

49. UNAIDS, "2008–2009 Unified Budget and Workplan: Technical Supplement," Twenty-Sixth Meeting of the UNAIDS Programme Coordinating Board, Geneva, 22–24 June 2010.

50. aids2031 Social Drivers Working Group, "Revolutionizing the AIDS Response: Enhancing Individual Resilience and Supporting AIDS Competent Communities," synthesis paper, aids2031 Social Drivers Working Group, Worcester, Mass., 2010.

51. Global HIV Prevention Working Group, *Global HIV Prevention Progress Report Card*, 2010.

# 3

# Using knowledge for a better future

This chapter envisages radical changes in our approach to AIDS programs and policies. Achievements will need to be measured with a longer-term perspective, focusing on outcomes rather than process. Uncomfortable topics that have long been avoided will need to be opened up and discussed frankly. These include the long-term future of AIDS treatment, or the equivocal evidence base of some of the programmatic strategies that the AIDS community has pursued. And all stakeholders will need to be more honest and critical when life-saving services are withheld from marginalized communities that need them the most, whether due to funding problems or through apathy, discomfort, or malice.

Much has been achieved in the planning and implementation of sound AIDS programs. As of mid-2010, an estimated 5 million people in low- and middle-income countries were receiving antiretroviral therapy, a 12-fold increase over numbers in 2002. Globally, the annual number of new infections has modestly declined.

Yet it is fair to ask whether the global AIDS effort has always achieved good value for its money. Despite a more than 53-fold increase in AIDS funding in barely over a decade, the epidemic continues to outpace the rate at which programs are delivering.

A review of spending patterns in sub-Saharan Africa suggests that knowledge is often poorly used to guide AIDS efforts (see Figure 3.1). For example, Botswana, Lesotho, and Swaziland are located in the same subregion and have comparable HIV prevalence, yet they devote markedly different shares of total AIDS spending to prevention, treatment, and orphan support programs. Rwanda, Congo, and Côte d'Ivoire have similar HIV prevalence, but administrative costs consume nearly four times as much in Rwanda and Congo. Settings undoubtedly matter, and there are legitimate reasons for divergent national approaches to

AIDS. But the sharp differences in resource allocations among similarly situated countries also suggest a lack of strategic rigor.

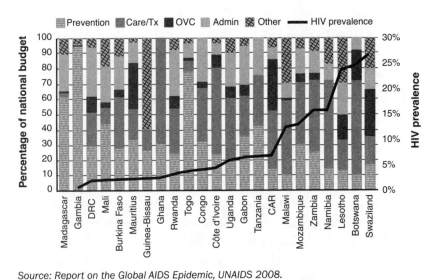

Source: Report on the Global AIDS Epidemic, UNAIDS 2008.

Figure 3.1  Allocation of HIV/AIDS resources in sub-Saharan Africa.

These failures are especially noteworthy with respect to HIV prevention. Although substantial evidence supports the need for strategies to prevent HIV, national strategies often ignore the evidence or use it poorly. Evidence on whether prevention programs are having any impact is typically lacking, and monitoring efforts generally focus more on counting the number of people who receive services than on measuring actual outcomes.

Prevention efforts also illustrate an additional impediment to success. Even when programs have been rationally planned, sound policies do not often accompany them. One out of three countries lacks a law barring HIV-based discrimination, dozens of countries have enacted punitive laws targeting people living with HIV, and the overwhelming majority of countries have laws in place that reinforce the marginalization of populations most at risk. Many national policies and programs leave HIV stigma largely unaddressed, and few countries in sub-Saharan Africa have implemented comprehensive policy frameworks to promote gender equity and empower women and girls.

This chapter proposes new ways of planning, implementing, and evaluating AIDS programs and policies. It proposes a much closer link between knowledge generation and actual programs and policies. It also argues for a shift in paradigms to generate a steady transition of programmatic authority from international agencies to countries and communities, with plans in place to build the local capacity that will be required to respond to AIDS over the long term.

## Participation at all levels

Some of the most enduring programmatic successes in the history of the AIDS epidemic have emerged not from peer-reviewed journals, public health agencies, or consultations of technical experts, but rather from the communities most affected by AIDS. In the early stages of the epidemic, even before epidemiologists had definitively characterized the modes of HIV transmission, urban gay communities in high-income countries were undertaking grassroots efforts to promote "safe sex." In the U.S., the emergence of innovative community mobilization strategies in the gay community—reflected, for example, in the risqué but highly effective "Safer Sex Comix" produced by New York City's Gay Men's Health Crisis—predated the creation of public sector funding streams to support HIV prevention for gay and bisexual men.

This same courage and creativity on the part of affected communities has been evident in regions across the world. For example, a robust civil society movement to combat AIDS arose in Uganda and was universally regarded as a critical element in that country's successful effort to lower HIV infection rates. That movement was tied not to government mandate, but to such people as the founders of The AIDS Service Organization (TASO). Despite their marginalization in most societies and the institutionalized discrimination they often face, networks of sex workers and drug users have also pioneered prevention models that are now recognized as the gold standard for programs. The worldwide expansion of treatment access would probably never have occurred without the relentless efforts of treatment activists. AIDS programs need to ensure that communities have the resources they need to innovate, and their insights need to serve as a cornerstone of program design.

To achieve the aids2031 vision of a transformed fight against AIDS, it will be necessary to harness, strengthen, and sustain the passion, intelligence, and commitment that have characterized grassroots leadership

on AIDS to date. Community leadership needs to be celebrated, culti-
vated, and sustained for decades to come. Without a true partnership
across governments, affected communities, and diverse nongovernmental actors, programs will not be optimally effective, their reach will
be inadequate, and any gains they achieve are unlikely to be maintained
over time.

## Designing more effective programs to prevent HIV transmission

If the picture of AIDS is to be substantially more favorable in 2031 than
it is today, markedly greater progress is required in preventing new infections. Given the rate at which the pandemic itself is outpacing programmatic scale-up, incremental improvements will not suffice. Prevention
programs will need to have radically greater impact in the coming years
if a sense of achievement, rather than despair, is to mark the 50th
anniversary of the first report of AIDS.

Ensuring success over the next generation demands immediate steps
to reverse the historic underinvestment in HIV prevention. Current
spending on HIV prevention is highly inadequate to provide prevention
services to those who need them.[1] In addition, new approaches to the
design and delivery of prevention programs are needed to maximize the
impact of available tools.

### Focusing prevention programs where they are most needed

A first order of business in expanding and improving HIV prevention is
to focus programs where they are most needed, using locally specific
knowledge to determine prevention strategies, priorities, and resource
allocations. National governments will continue to have a key role in allocating resources for HIV prevention, but district and local levels need to
assume more authority in planning and setting priorities, to ensure the
local relevance of prevention strategies. Recent estimates of HIV incidence by modes of transmission in different countries have demonstrated that national prevention priorities often bear little resemblance
to the actual cutting edge of the epidemics. For example, in Swaziland,

where people over 25 years of age account for two out of three new infections, few prevention programs focus on older adults.[2] Uganda's prevention strategies are heavily weighted toward young people, yet high rates of new infection among adults in monogamous and steady partnerships represent 43% of new infections.[3]

Recent analyses have particularly noted the tendency for decision-makers to focus insufficient effort on key populations at elevated risk, especially men who have sex with men, people who inject drugs, and sex workers and their clients. Even though men who have sex with men account for 15% or more of incident infections in Kenya and parts of West Africa, prevention programs for this population are virtually nonexistent in such settings.[4]

The recommendation to use epidemiological and sociological data to determine prevention priorities is not meant to suggest that there is a magic formula for allocating finite resources. In particular, a low percentage of new infections in a particular population may indicate that existing prevention strategies are working, not that the population has no need for services. Indeed, a perfect recipe for a resurgence of AIDS in a vulnerable population is the premature withdrawal of prevention services. As with treatment, prevention of infection is a lifelong undertaking that requires reinforcement and adaptation over time.

## Eliminating mother-to-child transmission of HIV

While taking steps to strengthen and reform our approach to HIV prevention, we should aggressively work to grasp other prevention opportunities that are staring us in the face.

In 2009, 370,000 children were newly infected with HIV, the vast majority of them during pregnancy or delivery, or as a result of breastfeeding.[5] Yet the tools exist to move rapidly toward eliminating mother-to-child HIV transmission. These tools include primary HIV prevention for reproductive-age women, family planning services for women living with HIV, HIV testing and counseling in antenatal settings, and timely administration of antiretrovirals to mother and newborn.

High-income countries are within sight of virtually eliminating mother-to-child transmission. For example, in the U.S., rigorous implementation of the recommended package of prevention services has reduced the proportion of infants born to HIV-positive mothers who themselves become infected from 25% to 2%.[6]

Despite a slow start in implementing prevention programs in antenatal settings, many countries have substantially expanded access to these services in recent years. In Southern Africa, coverage for services to prevent mother-to-child transmission reached 53% as of December 2009.[7] In sub-Saharan Africa, home to 90% of all children who become infected, two out of three countries have national plans for scaling up prevention of mother-to-child transmission, with associated national targets.[8]

Countries that have brought prevention of mother-to-child transmission to scale are approaching effectiveness levels seen in high-income countries. In Botswana, Rwanda, and Zambia, the percentage of infants born to HIV-infected mothers who become infected has fallen by 80% or more over the last six years.[9]

Despite encouraging overall trends, far too few HIV-positive pregnant women have access to the package of highly-effective, well-characterized services to reduce the risk of HIV transmission to their newborn. With such a clear and potent set of tools, urgent action is needed to close remaining gaps.

### Finding synergies in HIV prevention

A number of factors influence HIV transmission: individual norms and desires, physical environments that encourage or impede risk, the dynamics of sexual coupling, access to prevention tools and services, prevailing social norms regarding human sexuality or drug use, gender norms and other broad societal beliefs, and policy frameworks. Effective disease prevention must not only address the individual, but also focus on the social ecology in which individual decisions are made.

Virtually every major public health success has been built on a combination of behavioral, biomedical, social, and structural approaches. As just one example, consider efforts to reduce the harmful effects of tobacco use. These have included individual smoking cessation interventions; biomedical tools to alleviate nicotine dependence; marketing strategies to forge social norms that discourage smoking; and a broad range of policy reforms, including smoking bans in public places, litigation against tobacco companies, and punitive taxation schemes for tobacco products.

A comprehensive sustainable approach to HIV prevention must combine behavioral strategies that promote risk reduction, biomedical strategies that reduce the biological likelihood that risky behavior will result in HIV transmission, and social and structural interventions that address the environmental factors that either impede or support risk reduction.

HIV prevention efforts to date have been overwhelmingly oriented toward a narrow spectrum of individual behavioral interventions, complemented by a handful of biomedical tools (such as antiretrovirals to prevent mother-to-child transmission and adult male circumcision) and minimal support for broad-based social or structural approaches. If one reviews most national HIV prevention plans or grant proposals to the Global Fund to Fight AIDS, Tuberculosis and Malaria, prevention efforts are typically described as a disconnected set of strategies.[10] Frequently, no overarching goals for reducing new infections or changing behaviors are articulated, and potential synergies between prevention and treatment programs are not well examined, at a time when, for example, evidence is growing that antiretroviral therapy can play an important role in reducing HIV transmission.

This haphazard approach to program planning and design for HIV prevention contrasts sharply with the approach taken for treatment programs. In the case of treatment, the vast majority of countries have established clear national targets for treatment scale-up. Systemic barriers to treatment expansion (such as weak procurement and supply management, or poorly developed regulatory systems) have been identified, mapped, and addressed. Rigorous monitoring and evaluation systems have been put in place, and strategic efforts have been implemented to overcome capacity constraints on treatment scale-up. From national policymakers to clinicians, to community leaders, intensive efforts have focused on ways to increase knowledge of HIV status, bring patients into health systems, and help individuals adhere to treatment. Where major gaps have been noted, such as the lack of robust pharmacovigilance systems or inadequate capacity to monitor the emergence of drug resistance, international players have worked with countries to put new systems in place to fill these gaps.

Of course, important differences exist between HIV prevention and treatment. Yet the chasm between our approach to treatment and our efforts to develop, implement, and monitor prevention programs and policies is wide, indeed.

Any sound prevention plan must combine diverse programmatic elements based on documented needs, define how success will be achieved, and articulate a robust approach to monitoring and evaluation to see whether the program is achieving its desired impact. This approach is not a new idea, yet it is striking how seldom this common-sense approach has been put into practice.

To improve the effectiveness and sustainability of AIDS efforts, program planners need to pay greater attention to potential synergies between different components of "combination prevention". For example, although HIV testing and behavior change programs undoubtedly have a role to play in prevention efforts, they may have greater impact if they are combined and focused on particular populations. For example, consider serodiscordant couples, in which the uninfected partner stands a roughly 80% chance of becoming infected within a year if the infected partner is not treated.[11] In Kenya, an estimated 44% of married or cohabitating people living with HIV have partners who are uninfected.[12] HIV testing has not always had the desired prevention impact in every setting or for every population, yet overwhelming evidence points to the fact that testing couples together dramatically lowers the odds that the uninfected partner will contract HIV.[13] Focusing testing efforts on couples represents precisely the kind of high-impact strategy that long-term success on AIDS requires.

Other synergies are also apparent but often imperfectly captured. Harm-reduction programs for drug users, for instance, will be more effective if they are combined with policy changes or with outreach to reduce official harassment of program participants and service providers.

A redesigned response needs a longer-term horizon. For example, trial results demonstrating that adult male circumcision sharply reduces the risk of female-to-male sexual transmission have prompted countries with high HIV prevalence and infrequent rates of circumcision to begin designing programs to deliver safe circumcision services to adult males. By 2009, 13 countries in sub-Saharan Africa had developed or were in the process of developing operational plans to offer circumcision services to men. This is a healthy development, but a long-term planning perspective would supplement this adult-focused approach with planning and implementation of routine circumcision of male *neonates*. The reasons for a broader approach to circumcision scale-up are apparent if one thinks about the long term. In 2031, a male child born the year of this book's publication will be 21 years old. If current patterns of sexual behavior continue, the majority of this year's male infants will likely be sexually active, and a considerable percentage will already be infected with HIV. By planning *now* to minimize the epidemic's long-term burdens, programs will achieve a greater, more sustained effect.

## Knowledge alone is not enough—it needs to be used: the example of harm reduction

As in the case of preventing mother-to-child transmission, we know how to prevent people who inject drugs from becoming infected. HIV prevention for people who inject drugs involves a package of interventions, including access to sterile injecting equipment, counseling and behavior change interventions, basic health services, and access to drug substitution therapy or other forms of drug rehabilitation treatment.

Where this "harm reduction" package has been tried, the results have been remarkable. In New York City, the rate of new infections among people who inject drugs fell by nearly 80% between 1990 and 2002 following the implementation of harm-reduction programs.[14] This is not an isolated achievement, as virtually all studies throughout the world have demonstrated the public health and human benefits of harm reduction.[15]

Yet these highly effective approaches are far too seldom put into practice.[16] Although 148 countries have reported injecting drug use,[17] more than two-thirds of countries do not have a needle and syringe program; an even smaller proportion of countries offer opioid substitution therapy.[18]

These programmatic and policy deficits are not only counterproductive with respect to millions of people worldwide who are at risk of contracting HIV during injecting drug use, but they also risk allowing controllable epidemics to spread unnecessarily. Throughout much of Eastern Europe and Central Asia, for example, an epidemic that was once almost exclusively driven by transmission during injecting drug use is now increasingly characterized by substantial transmission to the sex partners of people who inject drugs.

Recognizing the widespread failure to use evidence to drive policies and programs around drug use, "The Vienna Declaration"[19] was launched at the 18th International AIDS Conference (2010) to mobilize governments and international organizations to take action at various levels, including to

- Undertake a transparent review of the effectiveness of current drug policies.

- Implement and evaluate a science-based public health approach to address the individual and community harms stemming from illicit drug use.

- Decriminalize drug users, scale up evidence-based drug dependence treatment options and abolish ineffective compulsory drug treatment centers that violate the Universal Declaration of Human Rights.

- Unequivocally endorse and scale up funding for the implementation of the comprehensive package of HIV interventions spelled out in the WHO, UNODC and UNAIDS Target Setting Guide.

- Meaningfully involve members of the affected community in developing, monitoring, and implementing services and policies that affect their lives.

### Addressing key social forces and structural factors

To address the long-wave phenomenon of HIV, greater efforts must focus on alleviating the social, political, economic, and environmental factors that increase vulnerability to infection.[20] Greater attention to social and structural factors acknowledges the reality that human behavior typically involves an interplay among individual factors (such as knowledge, motivation, skills, and opportunity), environmental factors (such as geographical settings or physical conditions that either promote or inhibit risk reduction), social factors (such as social norms, gender relations, and social capital), and structural factors (such as laws, policies, and economic or institutional arrangements that affect individual decision-making or access to prevention services).

Planning for social and structural components of combination prevention should be driven by the locally specific sociological assessments discussed in Chapter 2, "Generating knowledge for the future." Based on the articulated causal pathways in particular settings and evidence collected to test social science hypotheses, decision-makers will be better equipped to move from a short-term, emergency-style response to one that is more sustainable and enduring. For example, depending on documented needs and national circumstances, a comprehensive strategy to prevent transmission during sex work combines typical short-term approaches (for example, condom distribution, community mobilization, and increased health service access) with longer-term strategies, such as legal reform, efforts to alter gender norms, creation of employment alternatives to sex work, and efforts to deter sex tourism.

Anti-stigma programming offers another example of how a strategically designed program of combination prevention might differ from current approaches. Stigma is almost universally cited as a major impediment to service scale-up and utilization, and most countries include anti-stigma language in their national AIDS plans. But national plans seldom articulate the specific elements and desired intermediate and ultimate outcomes of anti-stigma strategies, and often do not specifically budget for stigma programming. As Figure 3.2 describes, anti-stigma efforts are seldom monitored against specific performance indicators. Moreover, many countries undermine the success of even haphazardly assembled anti-stigma efforts through policy frameworks that often reinforce and perpetuate stigma. As recommended by the aids2031 Programmatic Response Working Group, every country should have specific budgets, work plans, and performance targets, to implement strategies to alleviate stigma and prevent discrimination.

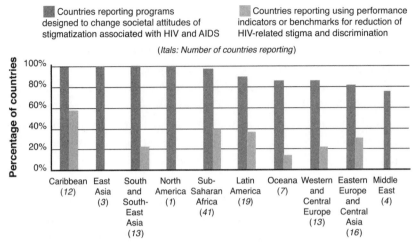

Adapted from UNAIDS Country Progress Reports 2008.

Figure 3.2   Percentage of countries (by region) reporting programs designed to change societal attitudes of stigmatization associated with HIV with and without impact indicators.

To develop, implement, and monitor strategies to address social and structural drivers, decision-makers should undertake an evidence-guided, multistep process. To take the example of addressing a concentrated epidemic characterized by substantial HIV transmission among sex workers, decision-makers should consider the following:

1. Identify the target populations and locations for the intervention.

2. Identify the key behavioural patterns and drivers of behavioural patterns for the target population (for example, nature of sex work, primary motivations for becoming involved in sex work, applicable legal frameworks, crucial power-brokers, role of law enforcement).

3. Choose the level of structural intervention (for example, legal reform, community empowerment, interventions to reduce police violence, and so on).

4. Describe planned or potential changes and outcomes as a result of the structural intervention (for instance, increase in condom use, formation of sex worker networks, reductions in reported violence against sex workers, changes in police attitudes).

**5.** Design the intervention (for example, enhanced opportunities for self-help among sex worker networks, condom access and promotion, anti-violence programs, mechanisms for legal recourse, advocacy for decriminalization, income-generating alternatives, and so on).

**6.** Implement, monitor, evaluate, and feed results into program design and implementation.

This approach is potentially useful in addressing epidemic drivers in broadly diverse settings. For example, where locally specific analyses suggest that gender inequality is increasing the vulnerability of women and girls, program planners may reasonably conclude that strategies to promote gender equality and empower women represent essential components of a comprehensive prevention response. Depending on available data, such approaches could include microfinance initiatives, legal recognition of property and inheritance rights, and other programs to increase women's earning potential and economic independence; legal reform to outlaw violence against women; and social change processes to alter gender norms, with a particular focus on changing the attitudes and behavior of men and boys.

More focused policy reforms may also contribute to risk reduction or improve the effectiveness of prevention efforts. For example, Thailand, Cambodia, and other countries have used regulatory powers to increase condom use in brothels, leading to dramatic reductions in HIV transmission. Similarly, changing laws to allow pharmacies to sell syringes has been shown to reduce needle sharing and the prevalence of used syringes in public places.[21] Analyses also suggest that implementing supervised injection facilities for drug users is a cost-effective strategy to reduce the sharing of needles and to link people who use drugs into service systems.[22]

As with other components of combination HIV prevention, social or structural interventions need to be carefully planned and strategically incorporated as elements of a comprehensive approach to AIDS. Informed by a rigorous analysis of relevant social dynamics, processes, and contexts, structural approaches should be articulated with clear causal pathways, incorporating specific, well-designed indicators for monitoring progress. Above all, social and structural interventions cannot be viewed as a short-term fix in the way that behavioral interventions have often erroneously been regarded. Structural approaches need to be linked to budget lines that are sufficient to support project cycles of 5 to 15 years or more.

### Sustaining HIV/AIDS treatment

The AIDS field has drawn inspiration and a sense of accomplishment from the steadily rising rates of treatment coverage in recent years. These achievements are real, but they also mask a more disturbing reality. With at least two million people dying of AIDS-related causes each year, the reality is that coverage levels are artificially inflated by our failure to deliver life-prolonging treatments to those who need them. The continuing high death toll—most of it preventable—effectively reduces the treatment denominator, partially obscuring the degree to which the goal of universal access to treatment remains unfulfilled.

We can and must do better. It will not be possible to achieve ambitious aims for significantly reducing the prevalence of HIV by 2031 without ensuring treatment to those who need it. Despite the considerable progress that has been made in scaling up treatment programs, most people who need antiretroviral therapy currently lack access.[23] Although the scale-up of treatment programs has arguably been guided somewhat more by evidence than corresponding efforts to prevent new infections, important challenges that threaten the long-term viability of treatment efforts remain. In particular, changes in program design and delivery strategies are needed to maximize the efficiency and sustainability of HIV treatment. A new initiative to maximize the use of antiretroviral therapy, called "Treatment 2.0," advocates for "better combination of treatment regimens, cheaper and simpler diagnostic tools, and community-led approaches to delivery."[24]

### Minimizing treatment costs

Many countries are missing opportunities to lower the costs of commodities used for treatment, even in an era when generic competition, agreements with pharmaceutical companies, and the support of such organizations as UNITAID and the Clinton Foundation are helping reduce the costs of drugs and diagnostics. According to one aids2031 analysis, there is considerable variation across countries in the prices paid for the same antiretroviral drugs, even among countries of a similar socioeconomic status.[25]

To maximize the long-term impact of finite funding, it is essential that national decision-makers, program implementers, and international donors collaborate to make sure that procurement of medicines and other health commodities is as efficient as possible. WHO's AIDS Medicines and Diagnostic Service provides up-to-date guidance on the most recent purchase prices for antiretrovirals and other technologies, but it is evident that many stakeholders are not availing themselves of this resource. Incentives should be built into AIDS financing schemes to ensure that purchases reflect the best value for money (an issue that Chapter 4, "Financing AIDS programs over the next generation," addresses in greater detail).

Additional work is also needed by organizations such as the Clinton Health Access Initiative (CHAI), formerly called the Clinton Global HIV/AIDS Initiative, to further reduce the costs of producing key commodities. Mechanisms to lower commodity costs include technology transfer, facilitated negotiations between different players in the production chain, volume guarantees, pooled purchasing, and even support for minority producers to guarantee competition.

### Maximizing treatment gains: a new approach to HIV treatment

In 2008, all but 2% of patients on antiretroviral therapy in low- and middle-income countries were receiving first-line regimens.[26] However, the growth over time in demand for second-line regimens is inevitable, as drug resistance develops. While the prices of second-line regimens have fallen somewhat in recent years, they remain several times more costly than first-line regimens.[27] Sustaining HIV treatment access requires concerted global efforts to maximize the time patients are able to sustain first-line regimens and to lower the prices of the second-line regimens that will be required in future years.

Clear evidence does not exist regarding which affordable regimen will last the longest and thereby preserve treatment gains. As Figure 3.3 illustrates, developing countries have opted for a wide range of first-line regimens. These patterns strongly suggest that many countries are placing patients on suboptimal regimens that invite emergence of resistance, increased patient drop-out due to toxicities, and expedited the need for more costly second- and third-line regimens.

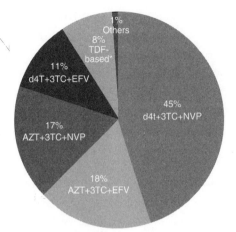

*TDF-based regimens: TDF+FTC=NVP (3%), TDF+FTC+EFV (3%), TDF+3TC+NVP (1%), and TDF+3TC+EFV (1%)*

Source: WHO Towards universal access report 2009.

Figure 3.3    Main first-line antiretroviral regimens used among 2.4 million adults in 36 low- and middle-income countries (December 2008).

This is a critical research priority to enable program planners in low-income countries to use limited resources to maximize individual health and longevity.

Especially in low-income countries where large proportions of the population lack access to the most rudimentary health services, existing second-line regimens likely will remain unaffordable. With competing health priorities, some health systems may decide that it is impossible to deliver second-line regimens through the public sector. Although it is the conclusion of aids2031 that sufficient resources exist to ensure access to high-quality HIV prevention and treatment services over the next generation (a proposition more fully explained in Chapter 4), prioritizing among competing health needs is the job of national governments and country-level stakeholders, not technical experts at the global level.

Regardless of a country's decision regarding access to second- and third-line drugs, it is evident that selecting the longest-lasting first-line regimen is an urgent necessity. Not only will optimally effective first-line regimens delay or avert the much higher costs associated with second-line drugs, but in settings in which second-line therapy is not available, they will help keep patients alive and healthy until future research breakthroughs can deliver additional, more affordable options for resource-limited settings.

*Enhancing treatment adherence*

To date, treatment scale-up has been based almost exclusively on a tradi-tional medical model. Clinicians aim to prescribe effective regimens and apply the best diagnostic tools at their disposal. From a programmatic standpoint, monitoring has focused primarily on the number of individu-als receiving antiretroviral treatment.

Although all these factors are critically important, they are insuffi-cient on their own to maximize and sustain medical outcomes. HIV treat-ment demands especially high levels of treatment adherence, with suboptimal adherence representing the primary cause of treatment fail-ure. With future access to second-line regimens uncertain in many of the countries most heavily affected by the pandemic, maximizing treatment adherence is vital to long-term sustainability of treatment gains.

Several case studies have identified patient support strategies that appear to increase treatment adherence, such as treatment "buddies" who work with patients to address impediments to adherence.[28] However, these initiatives are infrequently incorporated in clinical set-tings. Some studies have found antiretroviral adherence rates in develop-ing countries to be equal to or superior to those documented in high-income settings,[29] while other field-based studies have detected sig-nificantly less favorable adherence levels.[30]

Funders and program implementers urgently need to prioritize adherence support services as essential components of HIV treatment programs. These services are vital to optimizing treatment gains over the long term.

## An enabling environment for a sound programmatic response

For a disease so closely linked to social context, with burdens dispropor-tionately visited on key marginalized populations, the importance of a sound and supportive legal framework is evident. Unfortunately, prob-lematic legal frameworks continue to undermine rights-based approaches to AIDS and deter individuals most at risk from using needed services. This problem is not new, but has instead been a fundamental weakness of AIDS efforts from the epidemic's beginning. As we look toward the next generation and the challenge of sustaining an optimally effective AIDS response, significantly stronger global solidarity and greater political courage are needed to remove legal and policy barriers.

Institutionalized discrimination reinforces the social marginalization of the groups at highest risk of infection. Despite the strategic imperative of delivering evidence-informed prevention services to these vulnerable groups who are being devastated by the epidemic, countries too often erect immense barriers to sound, rights-based programming. Among 92 countries reporting data to WHO and its partners in 2009, only 30 countries permitted needle and syringe exchange programs, and opioid substitution therapy was available in only 26 countries.[31] As Figure 3.4 illustrates, such barriers are especially common in Eastern Europe and Central Asia, where AIDS epidemics have historically been driven by drug-related transmission. More than 50 countries impose coercive or compulsory treatment or the death penalty for people convicted of drug offenses,[32] a legal approach that inevitably drives drug users away from services. Seventy-six countries (most of them low- and middle-income countries) criminalize sexual relations between members of the same sex, including seven that provide for the death penalty.[33] And laws in at least 110 countries make it illegal to offer or solicit sex in exchange for money.[34]

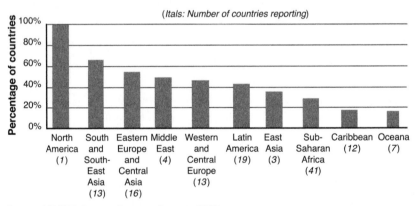

Source: UNAIDS Country Progress Reports 2008.

Figure 3.4    Percentage of countries reporting laws, regulations, or policies that present obstacles to services for injecting drug users.

A prerequisite to a sound AIDS response is for all countries to have in place a *minimum legal framework* to remove key legal and policy impediments. As recommended by the aids2031 Social Drivers Working Group,[35] this minimum legal standard includes the following:

- Decriminalize HIV status, transmission and exposure
- Decriminalize same-sex relationships/sexual practices

- Guarantee equal rights of people living with HIV
- Guarantee equal rights to men and women
- Eliminate laws that limit access to health services for marginalized populations, including sex workers, people in same-sex relationships, and drug users
- Decriminalize harm-reduction approaches for prevention of AIDS among those injecting drugs

If widespread enactment of this minimum framework is to come about, political decision-makers cannot be let off the hook for actions that undermine AIDS efforts. Global solidarity can, in fact, make a difference. In recent years, substantially greater attention has focused on the corrosive and discriminatory effects of national policies that restrict the ability of people living with HIV to cross national borders. Although such laws have been a feature of national AIDS policies since the pandemic's first decade, there are signs that this discriminatory edifice is beginning to crumble in the face of global criticism. Over the last two years, countries such as the Czech Republic, South Korea, and the U.S. have taken steps to remove such counterproductive policies.

### Building sustainable national and local capacity

Even while urging national ownership of the AIDS fight, the international community has largely sought to implement programs *for* developing countries. With a certainty that they know best how to respond to culturally distinct epidemics in countries they only partially understand, international donors and technical agencies have imposed theory-based approaches that inevitably obtain disappointing results. In the process, the international community has encouraged a dependence on external AIDS assistance in many countries that is wholly incompatible with the stated aim of national ownership.

Not only has this approach yielded unsatisfactory results in many cases, it is also a poor recipe for sustainability. Donors are often fickle when it comes to their commitments, and fashions in the international development field frequently change over time. Basing the long-term AIDS response on such a fragile base is dangerous.

This approach is also indicative of the kind of short-term perspective that has frequently characterized AIDS programs. In an emergency phase, getting programs up and running may indeed be the first priority.

Over the long term, however, it will be necessary to have strong, durable, self-replenishing mechanisms in place to support the generations-long struggle against AIDS.

### Transitioning to local control

A new approach is urgently needed to build the professional skills, managerial expertise, and analytical aptitude in the countries and communities that will grapple first-hand with AIDS for decades to come. As the aids2031 Programmatic Response Working Group recommends,[36] donors, developing country governments, and multilateral agencies should enforce a "new compact" for AIDS programs. Contracts for the delivery of AIDS programs in low- and middle-income countries should include clear, unambiguous milestones for building capacity. Where appropriate, this would include creating physical infrastructure and providing on-the-job training and professional development for in-country staff to acquire the skills and experience they will need to administer the program over the long run. Not only will this approach promote a sustainable response, but it will also be more efficient, as financial requirements for international staff are nearly always greater than for local personnel.

Each contract of an international NGO should include an extensive handover period, governed by specific milestones, during which day-to-day management of the program is steadily transitioned to in-country staff—either to a local community organization or to a public agency. International NGOs must understand that their ability to compete for future contracts from donor agencies will depend on a strong track record of building sustainable capacity in countries. And donors need to commit to a continuation of funding for successful, locally managed programs.

### Strengthening national systems

Weak national infrastructure has repeatedly undermined the impact of AIDS programs. Special attention has recently focused on strengthening national health systems, but a durable AIDS effort depends on the capacity of other systems as well, including education, welfare, and justice.

Although strengthening health systems is important, both for AIDS programs and other health priorities, the emerging discourse poses the challenge as an either/or choice between focusing on disease-specific programs and strengthening broader health systems. This type of thinking would have made the historic successes of the AIDS response impossible to achieve. The AIDS response has actually been one of the drivers of the increases in Official Development Assistance (ODA) for health during the last decade. Sufficient funds exist to provide robust support for *both* scaled-up efforts to address priority diseases *and* actions to strengthen national health systems, as Chapter 4 discusses in greater detail. Careful planning is needed to ensure that the broader systemic benefits of AIDS funding are maximized.

To the extent that the world is serious about strengthening health systems, it will demonstrate seriousness in monitoring and evaluating such initiatives. Initiatives for strengthening health systems should have clearly defined, transparent impact indicators in place, with regular reporting of results.

### Building AIDS-resilient communities

If AIDS has taught us one lesson, it is that communities, neighborhoods, and social networks have an enormous capacity to influence attitudes, drive change and innovation, and persuade decision-makers to pay attention to their needs. Indeed, one cannot understand AIDS without recognizing the potential for even the most marginalized and disempowered groups to create change. Building strong and enduring social capital in affected communities enables a robust community response to continually regenerate itself, a recipe for a more sustainable response.

Placing communities at the center of AIDS programs has important implications for programmatic effectiveness and sustainability. Programmatic efforts must be built on a strong community foundation to succeed over the long run.

Different types of social capital are important to a long-term AIDS strategy. *Bonding social capital* is the sociological term for the horizontal relationships between members of the same group that reinforce social solidarity. *Bridging social capital* refers to social interactions between different groups. *Linking social capital* connects groups of people across power hierarchies, connecting a less powerful social group with individuals or institutions that possess greater power and influence.

Bonding social capital is the glue that holds communities together. The power of group solidarity has helped make some of the most noteworthy community-centered HIV-prevention efforts so successful. Although it is most common to think of bonding social capital as a phenomenon that occurs within a geographic setting, the enormous expansion of international travel and communication has also given rise to emerging global "communities" centered on more shared characteristics than a common geographic setting. In contrast to the internal community dynamics that characterize bonding social capital, bridging social capital is defined by the external links that communities establish with other groups, potentially increasing the reach of community-generated strategies.

Linking social capital is the means by which a disempowered or disadvantaged group calls attention to its needs and persuades those with power to take action to benefit the community. Linkages between relatively disempowered communities and empowered actors with the ability to make or influence decision-making on public policy have played an important role in strengthening AIDS efforts.[37] The Treatment Action Campaign's successful drive to persuade the South African government to alter its misguided earlier approach to AIDS is an excellent example; although street protests, creative use of the media, and other political tools played a central role in bringing policy change, mobilizing partnership with diverse and influential actors was also key.[38]

Unfortunately, many trends in AIDS programming undermine the social capital on which a sustainable and vigorous response depends. Consider bonding social capital, for instance. The ideal end result of bonding social capital is the development of AIDS-resilient communities that collaborate to identify problems, agree on goals and priorities, and implement responses to address problems.[39] Characteristics of AIDS-resilient communities include open community dialogue, collaborative decision-making, and reliance on the community's strengths and resources to respond to problems.[40]

Unfortunately, key attributes of AIDS programming often weaken the social capital needed to produce AIDS-resilient communities. Community members may serve as key informants during a program's formative stage or sit on a program's consumer advisory board, for example, but far too often AIDS programs fail to encourage true community ownership or leadership.

## Respecting, preserving, and building social capital: the example of the HIV Collaborative Fund

Alternatives do exist to the traditional top-down approach whereby donors support civil society groups. The HIV Collaborative Fund, a project of the International Treatment Preparedness Coalition and Tides Foundation, supports AIDS treatment advocacy and education programs in four different regions.

Whereas donors seeking to support civil society groups often use technical review panels to score competitive applications from multiple applicants, the HIV Collaborative Fund convenes people living with HIV themselves to identify programmatic priorities in their own respective regions and to make funding decisions. To date, more than 100 community-based and regional organizations around the world are already involved in the HIV Collaborative Fund. The World Health Organization (WHO) selected the project as the sole funding recipient among treatment preparedness projects associated with the "3 by 5" initiative whose goal was to reach 3 million people living with HIV in low- and middle-income countries with antiretroviral treatment by the end of 2005.

The HIV Collaborative Fund model empowers communities to determine and own programmatic priorities in their own regions. The participatory process of selecting recipients also increases the likelihood that the civil society groups supported have a genuine connection to their communities and the ability to deliver results.

A separate set of problems impedes linking social capital. Often disparities in power are so vast that success stories such as the Treatment Access Campaign are unlikely ever to emerge without focused action. Yet given the documented power of community activism in the history of AIDS, it is remarkable how little funding is made available for civil society advocacy and watchdog functions.

### Investing in medical education

New cadres of AIDS-competent health professionals are needed to meet the long-term challenge AIDS poses in developing countries. Home to two-thirds of people living with HIV,[41] Southern Africa claims only 3% of the global health workforce.[42] Although task shifting, professional mentoring programs, and telemedicine techniques certainly help extend available human resources further than they would otherwise go, they cannot compensate over the next generation for an acute shortage of health workers needed to deliver essential treatment and care.

The prevailing scattershot approach to building durable human resources for health needs to be replaced by a much more ambitious, more strategic, adequately funded, decades-long initiative to build regional and national institutions capable of serving as a human resource pipeline for health systems. The well-documented "brain drain" of health personnel from poor countries to high-income countries has sometimes deterred donors from prioritizing investment in medical and public health education systems in low- and middle-income countries. This needs to change. Sustaining an effective AIDS response is simply not feasible without deploying substantial numbers of new health personnel in high-burden countries.

## Managing programs to enhance their long-term efficiency and effectiveness

In the transition from a singular emphasis on programmatic scale-up at any cost to an approach that focuses as much effort on maximizing quality, efficiency, and impact, programs need to be managed more rigorously for results.[43] This requires not only a new mentality among national governments, international donors, and program implementers, but also new skills in implementing organizations and access to user-friendly monitoring and management tools.

Recent experience has demonstrated the potential benefits of using business principles to guide programmatic scale-up. Created in 2002, the Avahan India AIDS Initiative, a joint collaboration between the Indian government and the Bill & Melinda Gates Foundation, undertook investigations and mapping to understand the target populations, conducted marketing research to determine their needs and preferences, established specific programmatic goals and benchmarks for program delivery and management, and implemented rigorous monitoring systems to determine whether such targets had been achieved.[44] These management practices and monitoring tools generated a high degree of confidence that Avahan had achieved the coverage levels it sought.

Prevention programs are not alone in the need to use business practices to guide programming and monitoring efforts. Treatment programs with high drop-out rates would also benefit from better monitoring systems. As Figure 3.5 illustrates, retention rates may be cause for concern; in Asia, for example, fewer than 60% of antiretroviral patients were still enrolled in treatment 48 months after starting therapy.

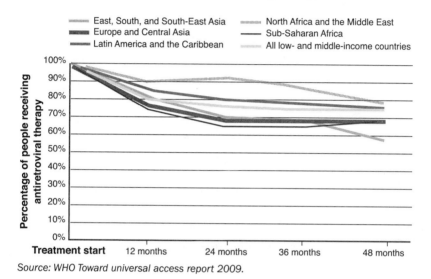

Source: WHO *Toward universal access report 2009.*

Figure 3.5  Trends in retention on antiretroviral therapy in low- and middle-income countries by regions (2008).

Patients may drop out of a treatment program for numerous possible reasons. A number of developing countries are using electronic medical record systems to better monitor and interpret data on patient retention, recognizing that such systems would more than pay for themselves by improving care coordination and preserving treatment gains.

### Ensuring quality

In a sustainable, long-term AIDS response, a commitment to scaling up must be matched by a commensurate dedication to the highest service quality. If a sex worker who seeks HIV counseling encounters social disapproval and hostility, for instance, the counseling program will not achieve its desired ends because the client is unlikely to continue seeking services at the site. As a recent analysis commissioned by UNAIDS advises, improving the quality of HIV-prevention programming involves a continuous loop of inspection, quality assurance, continuous quality improvement, and total quality management.[45] In short, quality control should continually inform service delivery and adaptation of programmatic efforts.

### Promoting the efficient use of finite resources

The paucity of data on AIDS service efficiency is striking, and the limited available evidence suggests that there is need for improvement. In a five-country study, the cost per client receiving HIV voluntary counseling and testing varied from US$3 to US$1,000, independent of the size of the program.[46]

A more rigorous approach to program management is required to increase efficiency. Mapping average unit costs for providers in a single country, it is possible to clearly differentiate the most efficient providers from the most inefficient.[47] Increased investment in operational research and more systematized service monitoring could lead to the establishment of baseline and target unit costs to guide program oversight.

Unfortunately, current ways of doing business are not always conducive to an efficient use of limited resources. For example, national governments and individual program managers are burdened with multiple, inconsistent donor-reporting requirements. In a typical low- or middle-income country that relies on international donor funding, multiple channels of aid from multiple sources require country officials to submit thousands of quarterly reports each year and host hundreds of site visits.[48] Instead of focusing on service delivery, country officials and program managers are drowning in paperwork. International donors have repeatedly pledged to streamline, harmonize, and align reporting requirements to support rather than undermine national strategies, including in the Paris Declaration on AIDS Effectiveness and in the broadly endorsed "Three Ones" approach to HIV programming. The "Three Ones" approach calls for 1) a single agreed-upon national AIDS framework that includes the work of all partners; 2) a single national AIDS coordinating body; and 3) a unified monitoring and evaluation system. Similar to repeated donor pledges to devote at least 0.7% of national wealth to international development assistance, the commitment to lessen reporting burdens and improve coordination in support of national plans has been honored only infrequently.[49] Taxpayers have every right to know that their country's spending on AIDS is being used for its intended purposes. However, the go-it-alone approach of individual donors with respect to reporting requirements reduces efficiency and diminishes the impact of AIDS spending. It also fails to promote accountability for *results* because few donor-reporting schemes collect information on the *impact* of their programs.

### Placing people living with HIV at the center of AIDS efforts

In the quest to sustain AIDS programs for the coming decades, it is obvious that no group in the world has a clearer stake in a sustainable response than people living with HIV. They are also ideally positioned to lead AIDS programmatic efforts and deliver essential services and support. Not only are people living with HIV uniquely qualified to provide information and support to their peers, but their encounters with service recipients are more likely to be informed by an intimate and nuanced understanding of the challenges these individuals face.

The absence of people living with HIV in programmatic efforts is especially acute in the prevention field. This is a curious shortcoming, in that individuals who are living with HIV often have an especially extensive understanding of the factors and circumstances that contribute to risky behavior and of the impediments that people face in avoiding transmission. For those who are uninfected, people living with HIV can also be the most compelling source of information and support for risk reduction.

AIDS programs, especially prevention programs, should undertake the massive hiring of people living with HIV. In addition to strengthening the relevance and cultural competence of AIDS services, this approach would help develop new cadres of AIDS leaders, support efforts to alleviate AIDS stigma, and provide economic opportunities for many people who currently lack them.

### Cultivating a new generation of AIDS leaders

To ensure that AIDS programs have the well-prepared leaders they need to thrive in future decades, focused efforts should be undertaken to cultivate, develop, and train a new generation of AIDS leaders. In 2009, aids2031 convened the second of two global Young Leaders Summits, where young people recommended steps to actively promote vibrant

youth leadership in AIDS programming. The young people at the summit endorsed a "5% for the Future Campaign," calling for at least 5% of
donor contributions to be allocated to youth-led organizations as a strategy to build sustainable AIDS leadership and encourage more youth-
relevant AIDS programs. They also recommended the establishment of
a Young Leadership Mentorship Hub to enable young leaders to
exchange ideas with, and learn from, more established leaders, including
decision-makers in the media and in the AIDS policy field.

## Political leadership for an effective response

This chapter has focused on translating knowledge into practice. With
better knowledge, the aids2031 Consortium contends, it will be possible
to achieve substantially more favorable results.

Too often lack of knowledge is not the only obstacle to a successful
AIDS response. Instead, it is the timidity of political leaders, who have
refused to put in place policies and strategies known to be effective.
Time after time, political leaders have heeded popular opinion more
than the evidence of what works.

Not only do political courage and commitment need to increase in
the coming years if the AIDS challenge is to be met, but a new type of
leadership also is required. For the typical political leader—focused on
the next election cycle or on time-limited work plans—the strategic horizon seldom extends beyond three or five years. However, mounting a
sustainable fight against the generations-long challenge of AIDS
requires continuity, long-term planning, and a broad national consensus.
Leaders need to summon the resources within themselves to make
sound decisions on AIDS, even when results may sometimes be apparent
only many years down the road.

Forging a durable national consensus on AIDS requires true leadership. Decision-makers need to avoid the temptation to pit groups against
one another or to pretend that AIDS is not a national priority because it
often primarily affects marginalized segments of society. Instead, leaders
need to lead, working to instill values of social solidarity, inclusion, and
compassion throughout society.

Over the next generation, the success or failure of the AIDS response will depend above all else on strong, enduring, evidence-based national leadership. But global solidarity also is critical to mobilizing needed resources and generating critical new knowledge and health tools. Preserving the potential of the AIDS response to inspire passion and goodwill will remain vital in the coming years.

With the looming challenge of climate change and a proliferating array of security issues, the global political agenda continues to expand. However, the pandemic is not going away. AIDS must remain high on the global political agenda—at the United Nations, among donor governments, and in key political groupings, such as the G-20.

In high-prevalence countries, AIDS needs to remain a paramount political issue in the coming decades, involving people from all walks of life in a common national endeavor. Senior political leaders and parliamentarians need to speak openly and often about their national epidemic. An annual parliamentary debate should be mandatory in high-burden countries, informed by a report on country progress, remaining gaps, and future action plans.

Building and maintaining national commitment will be especially challenging in countries with lower HIV prevalence. In these settings, AIDS will not be one of the foremost issues in the minds of political leaders, but it cannot be allowed to slip into oblivion, because this would risk an expansion of the epidemic. Here, regional groupings have a potentially important role to play in building solidarity and commitment for continuing vigilance.

Long-term success demands overcoming political impediments that have long frustrated AIDS efforts. In particular, political courage is required to prioritize sound, rights-based approaches with respect to marginalized populations that have too often been ignored, such as men who have sex with men, people who inject drugs, and sex workers. The difficulties in overcoming popular prejudices are real, but the history of AIDS teaches that genuine progress is achievable.

## Endnotes

1. Global HIV Prevention Working Group, *Global HIV Prevention Progress Report Card*, http://www.globalhivprevention.org (Accessed 7 August 2010).

2. Mngadi, S., N. Fraser, H. Mkhatshwa, T. Lapidos, et al., *Swaziland: HIV Prevention Response and Modes of Transmission Analysis* (Mbabane, Swaziland: National Emergency Response Council on HIV/AIDS, 2009).

3. Colvin, M., Gorgens-Albino, M., and Kasedde, S., "Analysis of HIV prevention responses and modes of HIV transmission: the UNAIDS-GAMET–supported synthesis process," 2008, http://www.unaidsrstesa.org/files/u1/analysis_hiv_prevention_resp onse_and_mot.pdf (Accessed August 1, 2010).

4. Gelmon, L., P. Kenya, F. Oguya, B. Cheluget, and G. Haile, *Kenya: HIV Prevention Response and Modes of Transmission Analysis* (Nairobi, Kenya: National AIDS Control Council, 2009); Bosu, W., K. Yeboah, G. Rangalyan, K. Atuahene, et al., *Modes of HIV Transmission in West Africa: Analysis of the Distribution of New HIV Infections in Ghana and Recommendations for Prevention* (Accra, Ghana: Ghana AIDS Commission, 2009).

5. UNAIDS, *AIDS Info: 2010 UNAIDS Reference Report* (Geneva: 2010).

6. Centers for Disease Control and Prevention, *Pregnancy and Childbirth*, 2007. Accessed 14 June 2010 at www.cdc.gov/hiv/top-ics/perinatal/index.htm.

7. UNAIDS and WHO. 2009. *Op cit*.

8. *Ibid.*

9. Botswana National AIDS Control Agency, *Progress Report of the National Response to the 2001 Declaration of Commitment on HIV and AIDS: Botswana Country Report 2010* (2010); Republic of Rwanda, *United Nations General Assembly Special Session on HIV and AIDS Country Progress Report: 2008-2009*, 2010. Accessed 26 May 2010 at http://data.unaids.org/pub/Report/2010/rwanda_2010_country_progress_report_en.pdf; Zambia Ministry of Health and National AIDS Council, *Zambia Country Report: Monitoring the Declaration of Commitment on HIV and AIDS and the Universal Access* (2010).

10. The Global Fund to Fight AIDS, Tuberculosis and Malaria, *Report of the Technical Review Panel and the Secretariat on Round 9 Proposals,* 20th Board Meeting, http://www.theglobal-fund.org/en/trp/reports (Accessed August 7, 2010).

11. Wawer, M. J., R. H. Grey, N. K. Sweankambo, D. Serwadda, et al., "Rates of HIV-1 Transmission Per Coital Act, by Stage of HIV-1 Infection, in Rakai, Uganda," *Journal of Infectious Disease* 191, no. 9 (2005): 1,403–1,409.

12. National AIDS and STI Control Program, *Kenya AIDS Indicator Survey 2007* (Nairobi: Kenya Ministry of Health, 2009). Accessed 23 June 2010 at www.usaid.gov/ke/ke.buproc/AIDS_Indicator_Survey_KEN_2007.pdf.

13. Allen, S., E. Karita, E. Chomba, D. L. Roth, et al., "Promotion of Couples' Voluntary Counseling and Testing in Individuals and Couples in Kenya, Tanzania, and Trinidad: A Randomized Trial," *BMC Public Health* 7 (2007): 349.

14. Des Jarlais, D. C., Perlis, T., Arasteh K, et al. "HIV incidence among injecting drug users in New York City, 1990 to 2002: Use of serologic test algorithm, to assess expansion of HIV prevention services," *Am J Public Health* 95 (2005): 1439–1444.

15. WHO, UNICEF, and UNAIDS, *Towards Universal Access: Scaling Up Priority HIV/AIDS Interventions in the Health Sector* (Geneva: World Health Organization, 2009).

16. Strathdee, S. A., Hallett, T. B., Bobrova, N., et al. "HIV and risk environments for injecting drug users: the past, present and future," *The Lancet* 376 (2010): 268–284.

17. Mathers, B. M., L. Degenhardt, A. Hammad, L. Wiessing, et al., "Global Epidemiology of Injecting Drug Use and HIV Among People Who Inject Drugs: A Systematic Review," *The Lancet* 372, no. 9,651 (2008): 1,733–1,745.

18. WHO, UNICEF, and UNAIDS. 2009. *Op cit.*

19. http://www.viennadeclaration.com/the-declaration.html (Accessed on August 5, 2010).

20. Rao Gupta, G., and E. Weiss, "Gender and HIV: Reflecting Back, Moving Forward," in *HIV/AIDS: Global Frontiers in Prevention/Intervention,* edited by C. Pope, R. T. White, and R. Malow (New York: Rutledge, 2008).

21. Groseclose, S. L., B. Weinstein, T. S. Jones, L. A. Valleroy, L. J. Fehrs, and W. J. Kassler, "Impact of Increased Legal Access to Needles and Syringes on Practices of Injecting Drug Users and

Policy Officers—Connecticut, 1992–1993," *Journal of Acquired Immune Deficiency Syndromes Human Retroviral* 10, no. 1 (1995): 71–72.

22. Des Jarlais, D. C., K. Arasteh, and H. Hagan, "Evaluating Vancouver's Supervised Injection Facility: Data and Dollars, Symbols and Ethics," *Canadian Medical Association Journal* 179 (2008): 1,105–1,106.

23. WHO, UNICEF, and UNAIDS. 2009. *Op cit.*

24. *Is this the Future of Treatment? Imagine Treatment 2.0,* UNAIDS *OUTLOOK* Report 2 (2010): 46–53.

25. Wirtz, V., Forsythe, S., Valencia-Mendoza, A., et al, "Factors influencing global antiretroviral procurement prices," (2009) (Suppl 1):S6.

26. WHO, UNICEF, and UNAIDS. 2009. *Op cit.*

27. *Ibid.*

28. Behforouz, H. L., P. E. Farmer, and J. S. Mukherjee, "From Directly Observed Therapy to *Accompagnateurs*: Enhancing AIDS Treatment Outcomes in Haiti and in Boston," *Clinical Infectious Diseases* 38, no. S5 (2004): S429–S436.

29. Coetzee, D., K. Hildebrand, A. Boulle, G. Maartens, et al., "Outcomes After Two Years of Providing Antiretroviral Treatment in Khayelitsha, South Africa," *AIDS* 18, no. 6 (2004): 887–895; Katzenstein, D., M. Laga, and J. P. Moatti, "The Evaluation of the HIV/AIDS Drug Access Initiatives in Cote D'Ivoire, Senegal, and Uganda: How Access to Antiretroviral Treatment Can Become Feasible in Africa," *AIDS* 17, no. 3 (2003): S1–S4.

30. Nwauche, C. A., O. Erhabor, O. A. Ejele, and C. I. Akani, "Adherence to Antiretroviral Therapy Among HIV-Infected Subjects in a Resource-Limited Setting in the Niger Delta of Nigeria," *African Journal of Health Sciences* 13 (2006): 13–17.

31. WHO, UNICEF, and UNAIDS. 2009. *Op cit.*

32. International Planned Parenthood Federation, *Verdict on a Virus: Public Health, Human Rights, and Criminal Law,* 2008. Accessed 23 June 2010 at www.ippf.org/en/What-we-do/AIDS+and+HIV/Verdict+on+a+virus.htm.

33. Ottosson, D., *State-Sponsored Homophobia: A World Survey of Laws Prohibiting Same Sex Activity Between Consenting Adults* (Brussels: International Lesbian, Gay, Bisexual, Trans, and Intersex Association, 2009).

**34.** International Planned Parenthood Federation, *Verdict on a Virus: Public Health, Human Rights, and Criminal Law,* 2008. Accessed 23 June 2010 at www.ippf.org/en/What-we-do/AIDS+and+HIV/Verdict+on+a+virus.htm.

**35.** aids2031 Social Drivers Working Group, *Revolutionizing the AIDS Response: Enhancing Individual Resilience and Supporting AIDS Competent Communities* (Clark University, Worcester, Mass.: aids2031 Social Drivers Working Group, 2010).

**36.** Aids2031 Programmatic Working Group Report, "Making Choices, Embracing Complexity, Driving and Managing Change: The HIV Programmatic Response Over the Next Generation," http://www.aids2031.org/working-groups/programmatic-response (Accessed August 9, 2010).

**37.** De Waal, A., *AIDS and Power: Why There Is No Political Crisis—Yet* (London: Zed Books, 2006).

**38.** Grebe, E., "The Emergence of Effective 'AIDS Response Coalitions': A Comparison of Uganda and South Africa," presentation paper, aids2031 Mobilizing Social Capital in a World with AIDS Workshop, Salzburg, Austria, March 2009.

**39.** Fisher, W. F. and B. Thomas-Slayter, 2010 Report and Recommendations from the Workshop on Mobilizing Social Capital in a World with AIDS, March 2009 (Salzburg, Austria); Campbell, N., E. Murray, J. Darbyshire, J. Emery, et al., "Designing and Evaluating Complex Interventions to Improve Health Care," *British Medical Journal* 334, no. 7,591 (2007): 455–459; Campbell, C., Letting Them Die: Why HIV/AIDS Prevention Programs Fail (Indiana University Press, 2003); Lamboray, J., and S. M. Skevington, "Defining AIDS Competence: A Working Model for Practical Purposes," Journal of International Development 13, no. 4 (2001): 513–521.

**40.** Campbell, N., E. Murray, J. Darbyshire, J. Emery, et al., "Designing and Evaluating Complex Interventions to Improve Health Care," *British Medical Journal* 334, no. 7,591 (2007): 455–459.

**41.** UNAIDS and WHO, *AIDS Epidemic Update* (Geneva: UNAIDS and WHO, 2009).

**42.** World Health Organization, *The Global Health Report 2006: Working Together for Health* (Geneva: World Health Organization, 2006). Accessed 23 June 2010 at www.who.int/whr/2006/en/.

43. Bertozzi, S. M., M. Laga, S. Bautista-Arredondo, and A. Coutinho, "HIV Prevention 5: Making HIV Prevention Programs Work," *The Lancet* 372, no. 9,641: 831–844.

44. Verma, R., A. Shekhar, S. Khobragade, R. Adhikary, B. George, et al., "Scale-Up and Coverage of Avahan: A Large-Scale HIV-Prevention Program Among Female Sex Workers in Four Indian States," *Sexually Transmitted Infections* 86 (2010): 176–182.

45. Maguerez, G., and J. Ogden, *UNAIDS Discussion Document: Catalyzing Quality Improvements in HIV Prevention: Review of Current Practice, and Presentation of a New Approach* (Geneva: UNAIDS, 2009).

46. Marseille E., L. Dandona, N. Marshall, P. Gaist, S. Bautista-Arredondo, et al., "HIV Prevention Costs and Program Scale: Data from the PANCEA Project in Five Low and Middle-Income Countries," *BMC Health Services Research* 7 (2007): 108.

47. Bertozzi, S. M., M. Laga, S. Bautista-Arredondo, and A. Coutinho, "HIV Prevention 5: Making HIV Prevention Programs Work," *The Lancet* 372, no. 9,641 (2008): 831–844.

48. Birdsall, N., W. Savedoff, K. Vyborny, and A. Mahgoub, *Cash on Delivery: A New Approach to Foreign Aid with an Application to Primary Schooling* (Washington, D.C.: Center for Global Development, 2010).

49. *Ibid.*

# 4

## Financing AIDS programs over the next generation

The future of the AIDS pandemic will not be determined solely by money. But substantial financial resources *will* be needed if rates of new infections and AIDS deaths are to be sharply lower in 2031. Funding demands for the pandemic will grow, even under the most favorable of scenarios, but strategies exist today to lower the long-term cost trajectory if wise policy choices are made now.

The good news is that the level of resources that will be needed to achieve the desired outcomes in 2031 is achievable. The bad news is that some signs of donor fatigue are already evident. Too many donors and national governments seem to want to turn the page on AIDS, pretending that this is a problem that has been solved or that maintaining spending at or near current levels will be sufficient. As the preceding chapters make plain, nothing could be further from the truth. Indeed, failing to follow through on global commitments on AIDS will inevitably ensure that the pandemic will become more severe, more intractable, and far more costly in the future.

This chapter explores the future of AIDS financing. It projects future funding requirements to achieve optimal outcomes, using economic models to assess the long-term consequences of the decisions to be made in the coming years. The chapter also looks at options to diversify the sources of AIDS funding and to rationalize the financing of AIDS programs. It concludes with a series of urgent recommendations to improve the efficiency of AIDS spending, with the aim of ensuring that the world receives optimal value for the amounts it spends.

## Covering AIDS costs over the next generation

The unprecedented nature of the resources mobilized for AIDS has resulted in historic achievements—including the first-ever widespread introduction of chronic disease management for adults in low-income countries—but it has also led to a number of important misconceptions. One is that AIDS programs are so-called "luxury" items, while other priority health interventions are much less costly. Yet a comprehensive economic analysis by the aids2031 Costing and Financing Working Group confirms that the standard components of HIV prevention and treatment easily satisfy existing health standards for cost-effectiveness in low- and middle-income settings.[1] Another recent analysis commissioned by the Global Fund further suggests that the economic returns on AIDS treatment—through improved worker productivity and by averting future costs to care for children orphaned by the epidemic—may equal or outweigh the costs of treatment.[2]

Another popular notion is that AIDS has had more than enough attention and funding and that the AIDS field should now make room for other health priorities. Few in the AIDS world would argue against the need for massive and sustained investments to address a broad range of health issues in developing countries, but also recognize that, as Chapter 2, "Generating knowledge for the future," explains, failing to follow through in our AIDS efforts will have extraordinarily negative consequences, potentially resulting in the needless loss of tens of millions of lives. Even with a continuation of current trends, HIV incidence will likely increase in many countries. If an even less favorable scenario comes to pass, with actual cutbacks imposed on AIDS programs, a far more dire result is inevitable.

By contrast, as shown in Figure 4.1, intensifying the scale-up of existing strategies and combining them with a major focus on social and structural drivers could cut the annual number of new infections by more than half between now and 2031. And with the addition of new prevention technologies, even more favorable outcomes are achievable.

Figures 4.2 and 4.3 show how Asia and sub-Saharan Africa—with their high disease burden and significant population size, respectively—will account for the vast majority of AIDS spending needed over the next generation.

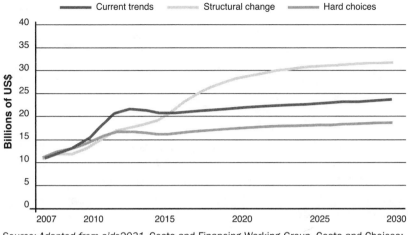

Source: Adapted from aids2031, Costs and Financing Working Group, Costs and Choices: Financing the Long-Term Fight Against AIDS (Washington, D.C.: Results for Development Institute, 2010).

Figure 4.1   Total projected annual AIDS resource requirements in low- and middle-income countries by scenario, 2007–2030.

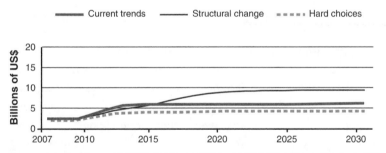

Source: Adapted from aids2031, Costs and Financing Working Group, Costs and Choices: Financing the Long-Term Fight Against AIDS (Washington, D.C.: Results for Development Institute, 2010).

Figure 4.2   Total projected annual AIDS resource requirements in Asia by scenario, 2007–2030.

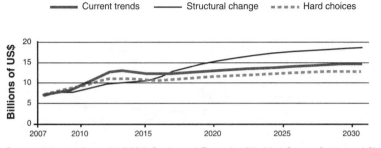

Source: *Adapted from aids2031*, Costs and Financing Working Group, Costs and Choices: Financing the Long-Term Fight Against AIDS *(Washington, D.C.: Results for Development Institute, 2010).*

Figure 4.3   Total projected AIDS resource requirements in sub-Saharan Africa by scenario, 2007–2030.

Recent funding trends foreshadow a potentially grim future for the pandemic. Even as long-term funding needs for HIV prevention and treatment continue to grow, the U.S. government has flattened funding for external AIDS assistance,[3] the Global Fund confronts a gap of several billion dollars between projected demands and available resources,[4] and surveys indicate that many developing countries expect to cap or reduce AIDS programs in the near future due to declining AIDS assistance.[5]

The sums required to finance AIDS programs over the next generation are certainly consequential. Mobilizing them demands both creativity and long-term political commitment. But in reality, they are eminently attainable. These costs could also be minimized through wise investments in AIDS research to develop new game-changing technologies, such as a preventive vaccine or more affordable treatments.

When one considers that AIDS funding means the difference between life and death for countless millions over the next generation, it is helpful to place these projected funding needs in context. The amount spent on AIDS programs in 2008 was less than half the amount the European Union spends each year on agricultural subsidies.[6] And closing the looming funding gap for AIDS would require only about 1% of annual global spending on armaments.[7]

Other optimistic reasons support the assertion that needed funding for AIDS can be obtained. Notwithstanding the historic global economic crisis from which many countries are only now beginning to emerge, economic projections commissioned by aids2031 indicate that the long-term trajectory for economic growth over the next generation remains favorable for both developed and developing countries. Moreover, as the subsequent section explains, numerous potential funding sources for AIDS programs have been either wholly untapped or inadequately utilized.

## Diversifying AIDS funding

Governments account for the lion's share of AIDS funding, either through domestic outlays, or through external assistance provided via bilateral or multilateral channels. This is entirely appropriate. As one of the most serious threats to public health and security, AIDS merits high-level support from governments across the globe.

But there is a pragmatic reason to diversify the funding base for AIDS. Government priorities often shift over time, sometimes quite suddenly. What is fashionable in development circles in one decade may be outmoded the next. However, radically lowering the number of new infections and AIDS deaths over the next generation requires sustained, high-level financial support. To optimize resource mobilization over the next generation, every available avenue for resources should be examined to establish durable, robust funding sources.

### Domestic AIDS financing

Notwithstanding the intensive focus on international sources of AIDS assistance, the largest share of AIDS funding comes from national governments and private households in affected countries. The capacity of affected countries to bear a comparable share of the financial burden as funding needs grow in the coming years depends on a combination of available funding of fiscal space and political commitment.

Not surprisingly, national wealth is directly correlated with per-capita health spending, as shown in Figure 4.4. As national economies grow and prosperity spreads, countries have a greater capacity to shoulder the financial burdens associated with domestic health needs. Given this relationship, it is apparent that pro-growth policies that are consistent with poverty reduction and greater income equality will play a potentially important role in expanding the resource base for AIDS and other health programs over the next generation.

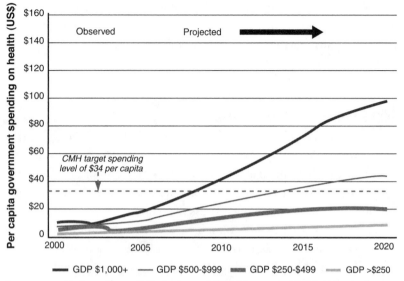

Source: van der Gaag, Hester V, Hecht R, et. al, Fiscal Space and Political Space for
Financing the Global AIDS Response to 2031. Aids2031 Working Paper #17.
(http://www.aids2031.org/working-groups/financing). Figure initially adapted from WHOSIS
and IMF World Economic Outlook database.

Figure 4.4   Current and projected government per-capita spending on health
by selected GDP levels.

In thinking about domestic financing capacity in the coming years, it
is useful to stratify countries according to per-capita wealth and projected
HIV prevalence. Middle-income countries with adult HIV prevalence
below 1%—including Brazil, China, India, Mexico, Russia, Thailand,
Ukraine, and Vietnam—should have the capacity to finance their AIDS
response on their own, without the need for external support. In
Southern Africa, a number of high-prevalence countries are much
wealthier than the regional average and are likely to experience impres-
sive economic growth between now and 2031; these countries—including
Botswana, Namibia, South Africa, and Swaziland—will be able to finance
a large share of their domestic AIDS needs but will almost certainly
require some external support over the next few years as they work to
expand treatment programs to meet the demand. A similar cost-sharing
arrangement between national governments and international donors
may be necessary for other low-income countries with significant epidemics, such as Burkina Faso, Cameroon, Ethiopia, and Nigeria.

Several other low-income countries with high HIV prevalence in
Africa will face considerable impediments to self-financing their AIDS

programs, even 10 or 20 years from now. Although countries such as Kenya, Mozambique, Uganda, and Zambia are projected to experience modest to strong economic growth over the next generation, the heavy financial burdens associated with their epidemics will perpetuate their current dependence on external donors.

These differences in economic prospects and AIDS burdens mean that different countries will—and should—be expected to adopt varying approaches to AIDS financing in the coming years. As Figure 4.5 illustrates, countries with very large populations, such as China and India, will have to spend large absolute amounts on AIDS. But AIDS spending in these countries—along with others with relatively low-level epidemics, such as Brazil, Thailand, Ukraine, and Vietnam—should consume only a tiny fraction of national wealth. These countries ought to be able to cover their AIDS costs with domestic funds. On the other hand, countries with a high AIDS burden and weaker economies, such as Mozambique and Zambia, will need to spend a much larger proportion of their national gross domestic product (GDP) on AIDS; but, because these countries have little prospect of fully covering their AIDS costs on their own, they will probably need to remain dependent on donors for years to come.

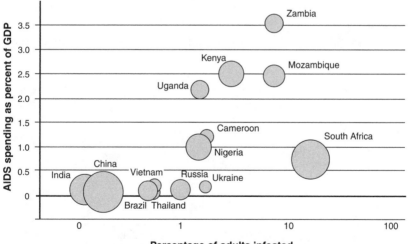

**Percentage of adults infected**

*Source: R. Hecht, et al., in* Health Affairs *28, no. 6: 1,591–1,605 (authors' calculations).*

Figure 4.5   Projected 2030 AIDS spending as a share of gross domestic product and adult HIV prevalence in selected countries. Note: Size of bubble is in proportion to each country's projected AIDS spending in a "rapid scale-up" scenario in billions of US$. Prevalence is calculated as the percentage of people age 15–49 infected with HIV.

These principles should be inscribed in the practices of international donors. Major donors such as the Global Fund and PEPFAR should prioritize funding for low-income, high-prevalence countries. By contrast, countries such as China and India, which are rapidly on their way to becoming major global economic powers, should transition as rapidly as possible to using purely domestic funds to pay for AIDS programs. Indeed, there ought to be a presumption against international funding for middle-income countries, although a certain degree of flexibility is merited for hyperendemic, middle-income countries such as Namibia and South Africa, which may require a degree of international support to finance their AIDS efforts.

### The future of international AIDS assistance

From a historical standpoint, the last decade has been a golden age of international development. Between 2002 and 2007, Official Development Assistance (ODA) nearly doubled. Development issues rocketed to the top of the global political agenda as the international community adopted a series of ambitious Millennium Development Goals to drive results between 2000 and 2015.

Ample room remains within the broader sphere of ODA to accommodate the funding needs required for AIDS over the next generation. As Figure 4.6 reveals, AIDS consumes only a small fraction of overall ODA.

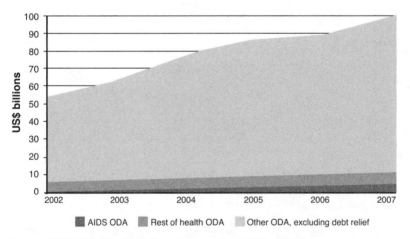

*Source: OECD/DAC CRS, 2009. Data extracted from OECD statistics.*

Figure 4.6   Total Overseas Development Assistance (ODA) from Development Assistance Committee members, 2002–2007.

Whether donors will capitalize on the available leeway to ensure robust financing is unclear. An analysis commissioned by aids2031 on the future of donor financing for AIDS identified several trends that are cause for concern.[8] In place of the disease-specific focus of many international health programs during the last decade, high-income countries are increasingly opting for sectoral approaches that seek to build broad-based systems rather than to address individual health conditions. Major European donors have long opted for this approach, and the U.S. government has also recently embraced this strategy. Moreover, the long-term financial burdens associated with helping developing countries cope with climate change will likely be considerable.

In addition, many high-income countries are experiencing structural budgetary constraints that call into question their likely commitment to international assistance in coming years. In particular, national populations in most industrialized countries are steadily aging. The proportion of national populations that will require social security benefits will continue to grow, but the base of younger workers who will have to pay the bill will likely be stable or decline in many countries. These pressures have been exacerbated by recent actions by high-income countries to respond to banking and currency crises, driving national debt loads to worrisome levels.

Although these trends are disquieting, others are potentially more positive. In particular, the emergence of new global and regional economic powers suggests that new AIDS donors could emerge. The generics pharmaceutical industry in India has already played an important role in facilitating improved antiretroviral access, and Brazil has pioneered numerous AIDS collaborations between developing countries. In particular, international AIDS leaders and diplomats from high-income countries should work to cultivate China as a provider of AIDS assistance to countries in Africa, where China has made significant infrastructure investments in recent years.

Above all, it is critical that the international donor community recognize its key role in influencing the future of AIDS. With many high-prevalence countries constrained in their capacity to finance domestic AIDS programs, closing the looming resource gap will only be possible if the international community increases its financial support. In particular, a number of major European countries lag in AIDS giving.

## Philanthropy and the future of AIDS

Although philanthropic foundations contribute a relatively small share of global AIDS spending (roughly 4% in 2008), these figures belie the role of foundations in driving innovation and raising awareness of emerging issues. For example, through its pioneering support for AIDS research and programmatic scale-up, the Bill & Melinda Gates Foundation occupies a dual role as intellectual and financial leader in the AIDS field. No other foundation matches the Gates Foundation's resources, but many others have played critical roles in supporting AIDS responses in developing countries. In particular, resource tracking indicates that foundations are especially vital sources of financial support for networks of people living with HIV.[9]

Yet despite the highly visible profile of leading foundations in the global AIDS effort, financial support for AIDS programs remains rare in the philanthropic world. Of the roughly 60,000 philanthropic foundations in the U.S., only about 80 contribute to international AIDS work.[10] The Gates Foundation alone accounts for 59% of all AIDS giving among U.S.-based foundations, and the top 10 philanthropic givers represent 82% of all AIDS funding from the foundation sector.[11] With a few exceptions, such as the Wellcome Trust and the King Baudouin Foundation, foundations from regions other than North America are not actively engaged in AIDS giving.

As Figure 4.7 illustrates, inflation-adjusted philanthropic giving in the U.S. has steadily risen over the last three decades. However, the unfortunate reality is that AIDS is not on the radar screen of most foundations. Many that prioritized AIDS programming in years past also are moving on to other issues.

These trends need to be reversed. Revision of national tax laws to encourage private philanthropic giving is urgently needed, especially in European countries, as one strategy to increase foundation support for AIDS and other global priorities

In addition, the AIDS field should do a better job of cultivating new philanthropic donors of the likes of Bill Gates, Warren Buffett, and George Soros. *Forbes* magazine reports that there are more than 40 billionaires under age 40. If each of the world's billionaires devoted 5% of his or her wealth each year to AIDS and other health challenges, the world could mobilize more than US$125 billion annually, five times the total current amount of ODA for health.

**US$ (billions)**

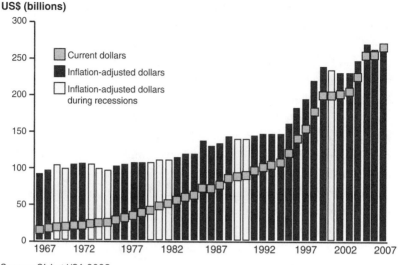

Source: *Giving USA 2008.*

Figure 4.7   Total giving in the U.S., 1967–2007. Recessions are in light green (1969–70, 1973–75, 1980, 1990–91, 2001).

## Innovative financing mechanisms for AIDS

Particular efforts are needed to develop additional, self-generating funding streams that do not rely on traditional sources of AIDS funding. AIDS has already pioneered numerous mechanisms to harness the purchasing power of consumers to mobilize resources for AIDS programs. Several of the world's leading product brands participate in the (RED) campaign, donating a percentage of their profits from sales of designated products to the Global Fund. Likewise, proceeds from the Viva Glam line of MAC cosmetics, which are promoted by leading celebrities, are directed toward the MAC AIDS Fund, which supports AIDS programs delivered by civil society groups in countries throughout the world.

The airline solidarity tax scheme, proposed by France and launched in 2006, generated US$718 million over a little more than two years for UNITAID, which procures essential AIDS medicines for programs in developing countries. UNITAID is being supported by the Millennium Foundation's "Massive Good" campaign, which offers purchasers of airline tickets the opportunity to make a US$2 contribution. More could be done to maximize the potential of the airline tax. In addition to expanding the airline tax, online travel agencies could permit purchasers of airline tickets to *opt out* instead of offering them the opportunity to opt in; an approach that would inevitably increase proceeds from the tax.

Other, even more ambitious financing mechanisms could be pursued. The Task Force on Innovative Financing for Health Systems has proposed a modest tax on currency transactions. The so-called Tobin tax, named after Nobel Prize laureate James Tobin, is believed to have the potential to generate between US$100 billion and US$300 billion annually. Those funds could be earmarked for essential global priorities, such as AIDS programs, health systems strengthening, infrastructure development, and climate change adaptation.

Other options also exist. The advance market commitment for new health innovations for developing countries (mentioned in Chapter 3, "Using knowledge for a better future," as a potential spur to greater industry investment in research and development) can also function as a financing mechanism, facilitating the rapid introduction of new HIV prevention technologies.

Renewed efforts are required to shore up the innovative mechanisms that have now become standard components of the international health architecture. Donors should replenish the coffers of the Global Fund, to ensure its ability to fund sound AIDS proposals from low-income countries. The Global Alliance for Vaccines and Immunization (GAVI) also requires support; and, although it has not yet played a role in the delivery of AIDS programs, it could be a potential vehicle for the swift introduction of future AIDS vaccines.

### Performance-based incentives

The Global Fund has distinguished itself by its commitment to results-based funding. Recipients that cannot account for funds or that fail to deliver as agreed risk losing funding. According to results from 130 countries that received Global Fund grants, 75% of countries met their performance targets; and 21% of countries failed to meet agreed targets but demonstrated the potential to achieve these targets in the future.[12]

However, the commitment to results-based management at the level of national programs needs to be extended to individual service settings. Studies suggest that linking provider pay to performance may help improve service quality and improve health outcomes.[13] Some countries have introduced incentives to encourage providers to become engaged in HIV care. For example, Rwanda has implemented a national outpatient incentive payment scheme that seeks to increase the frequency with which providers deliver specific HIV-related services.[14] Funders and program managers take this approach one step further, to reward providers that achieve excellent health outcomes for their patients, such as above-average patient retention, treatment adherence, and viral suppression.

### Incentives to promote efficiency and effectiveness

To date, a primary strategy for translating AIDS knowledge into action has involved the centralization of technical norm setting in Geneva. Through this approach, UNAIDS and the World Health Organization (WHO) review technical evidence and issue guidelines for AIDS programs. Although this approach has helped forge consensus on key technical issues, it has been much less successful at influencing actual programs and policies. For example, notwithstanding long-standing calls by global technical experts to give HIV prevention the priority it deserves, developing countries and donors currently spend US$2.50 on treatment for every US$1 allocated to prevention. And although international guidelines have long insisted on a rights-based approach to AIDS, one in three countries lacks a national law prohibiting HIV-based discrimination, and legal frameworks in dozens of countries institutionalize discrimination against the groups most affected by AIDS.

A new approach is in order that relies less on global cajoling and periodic technical missions and that focuses instead on creating financial incentives to encourage sound AIDS programming.

Conditioning future AIDS assistance on the effective targeting of programs for those who need them the most will go a long way toward improving the efficiency of AIDS programs. Unit costs for services also should decline as they are brought to scale.

But additional steps are required to ensure good value for each AIDS dollar. Startling variations in unit costs for standard AIDS services across different settings are common. Similarly situated countries also are often paying notably different prices for the same antiretroviral regimens. These and other inefficiencies should result in financial penalties.

To make the implementation of financial incentives feasible, significantly greater effort is needed to establish benchmarks for unit costs for well-characterized services (such as HIV testing and counseling, prevention of mother-to-child transmission, and administration of antiretroviral therapy). Reasonable variations should be allowed between low-volume and high-volume settings. Although cost overruns may be excused by exigent circumstances—such as civil unrest, natural disasters, or other unforeseen occurrences—providers should be expected to deliver services with basic efficiency.

Donor practices also need to change, to maximize efficiency. In 2005, bilateral donors and international agencies gathered in London for a landmark meeting titled, "The Global Response to AIDS: 'Making the Money Work', The Three Ones in Action" designed to improve the coordination and efficiency of AIDS giving. This meeting resulted in international agreement that donors should harmonize and align their efforts with national strategies and coordinating mechanisms. A global task team was established to improve coordination among multilateral agencies.[15]

Since this 2005 meeting, some improvements in donor coordination have been observed. In Malawi, Zimbabwe, and other countries, donors have pooled their giving in support of nationally-owned strategies and priorities. The working relationship between global agencies such as the Global Fund, UNAIDS, and the World Bank also has grown closer and better coordinated.

Yet substantial additional work is needed to improve the efficiency and effectiveness of donor programs. According to country reports submitted to UNAIDS in March 2010 regarding progress in the AIDS response, many countries continue to struggle to bring donor programs in line with national strategic plans. Far too many donors continue to pursue their own programmatic agenda, impose duplicative reporting requirements, and implement parallel monitoring systems. This urgently needs to change.

### Prioritizing HIV prevention

Developing countries are understandably receptive to external support for treatment programs. Every national government that is concerned about its citizens will look for all reasonable opportunities to prevent them from becoming ill or dying prematurely.

With the number of new infections vastly outpacing the rate of treatment scale-up, the queue for treatment grows longer by the day, giving rise to legitimate concerns about long-term treatment costs. As explained earlier in this chapter, aids2031 has concluded that ample financial resources exist to meet all AIDS-related prevention and treatment needs over the next generation. Nevertheless, the global community has a legitimate interest in ensuring that every attempt is made to limit the growth of future treatment demand by reducing the rate of new HIV infections. Accordingly, donors should more closely link future financial assistance for treatment programs to robust national support for evidence-informed prevention programs and policies. To be eligible for continued funding for treatment, countries should have prevention plans that set and meet specific measurable goals, articulate causal pathways and programmatic synergies, describe how prevention and treatment programs will be aligned and coordinated, and adhere to a rights-based approach. In an essay on long-term financing, Mead Over argues similarly that "new funding of AIDS treatment should be tightly linked to dramatically improved and transparently measured prevention of HIV infections."[16]

Specific incentives should focus on national targeting of HIV-prevention programs to those who need them the most. Especially (but not exclusively) in concentrated epidemics, financial incentives should be put in place to ensure sufficient programmatic emphasis on populations most at risk. Countries that maintain legal impediments to a rights-based approach to AIDS should incur penalties in the form of reduced access to external assistance.

As this chapter explains, the future capacity to finance a strong AIDS response over the next generation is not merely a matter of funding availability, but also one of political will. Not only do multiple strategies exist for mobilizing substantial new resources for AIDS, but options exist to make available funding go much further.

With respect to the aids2031 vision of achieving marked reductions in new infections and AIDS deaths by 2031, the relevant question is not *can* we do it, but *will* we?

## Endnotes

1. Demaria, L. M., S. A. Bautista-Arredondo, and O. Galárraga, *What Works to Prevent and Treat AIDS: A Review of Cost-Effectiveness Literature with a Long-Term Perspective,* working paper, Washington, D.C.: aids2031 Financing Working Group, 2009.

2. The Global Fund to Fight AIDS, Tuberculosis and Malaria, "Resource Scenarios 2011–2013: Funding the Global Fight Against HIV/AIDS, Tuberculosis, and Malaria," 2010. Accessed 26 May 2010 at www.theglobalfund.org/documents/replenishment/2010/Resource_Scenarios_en.pdf.

3. Henry J. Kaiser Family Foundation, *U.S. Federal Funding for HIV/AIDS: The President's Fiscal Year 2011 Budget Request,* 2010. Accessed 25 May 2010 at www.kff.org/hivaids/upload/7029-06.pdf.

4. The Global Fund to Fight AIDS, Tuberculosis and Malaria, "Resource Scenarios 2011–2013: Funding the Global Fight Against HIV/AIDS, Tuberculosis, and Malaria," 2010. Accessed 26 May 2010 at www.theglobalfund.org/documents/replenishment/2010/Resource_Scenarios_en.pdf.

5. UNAIDS and WHO, *AIDS Epidemic Update* (Geneva: UNAIDS and WHO, 2009).

6. European Commission, "Financial Programming and Budget," 2010. Accessed 7 June 2010 at http://ec.europa.eu/budget/budget_glance/what_for_en.htm.

7. Stockholm International Peace Research Institute, "Recent Trends in Military Expenditure," 2010. Accessed 26 May 2010 at www.sipri.org/research/armaments/milex/resultoutput/trends.

8. Schneider, K., and L. Garrett, *The Evolution and the Future of Donors Assistance for HIV/AIDS,* working paper, Washington, D.C.: aids2031 Financing Working Group, 2009.

9. Funders Concerned About AIDS, "U.S. Philanthropic Support to Address HIV/AIDS in 2008," 2009. Accessed 26 May 2010 at www.fcaaids.org/Portals/0/Uploads/Documents/Public/FCAART2009.pdf.

10. Kissane, R., *The Future Role of the Philanthropy Sector Fighting HIV/AIDS,* working paper, Washington, D.C.: aids2031 Financing Working Group, 2009.

11. Funders Concerned About AIDS, "U.S. Philanthropic Support to Address HIV/AIDS in 2008," 2009. Accessed 26 May 2010 at www.fcaaids.org/Portals/0/Uploads/Documents/Public/FCAART2009.pdf.

12. Low-Beer, D., H. Afkhami, R. Komatsu, P. Banati, M. Sempala, I. Katz, J. Cutler, P. Schumacher, T. Tran-Ba-Huy, and B. Schwartländer, "Making Performance-Based Funding Work for Health," *PLoS Medicine* 4, no. 8 (2007): e219.

13. Mills, E. J., J. B. Nachega, L. Buchan, J. Orbinski, et al., "Adherence to Antiretroviral Therapy in Sub-Saharan Africa and North America: A Meta-Analysis," *Journal of the American Medical Association* 296, no. 6 (2006): 679–690.

14. Management Sciences for Health, "Rwanda HIV/PBF Project," 2009. Accessed 7 June 2010 at www.msh.org/global-presence/rwanda-hiv-pbf-project.cfm.

15. http://www.unaids.org/en/CountryResponses/MakingTheMoneyWork/GTT/ (Accessed August 6, 2010).

16. Over, Mead, "The Global AIDS Transition" (May 2010). Accessed 30 September 2010 at www.cgdev.org/content/publications/details/1424143.

# 5

## The way forward: recommendations for long-term success

Sharply reducing the number of new infections and AIDS deaths by 2031 requires new ways of thinking about AIDS and responding to the challenges that the pandemic poses. It requires new prevention and treatment tools, sound policies to optimize the effectiveness of programs, innovative approaches to AIDS financing, the creation of strong and durable capacity in countries, transition from a focus on individuals to one that views communities as critical fulcrums for success, and management practices to maximize efficiency and effectiveness. The pandemic is not going away, but its magnitude and severity can be dramatically curtailed—if the global community brings the seriousness of purpose to this problem it deserves.

Given all that has been achieved in controlling the AIDS pandemic so far, we are not starting from square one. Rather, in some cases, we simply need to do more of what we have been doing, only better and more effectively. In other respects, though, achieving the aids2031 vision demands that we jettison old ways of doing business and venture out in new directions. Above all, we need to adopt a long-term perspective and recognize the pandemic for the generations-long challenge that it is.

As discussed in Chapter 2, "Generating knowledge for the future," certain breakthroughs—such as the development of a highly effective vaccine or the unexpected emergence of an affordable cure for AIDS—could dramatically alter the epidemic's course for the better. Many of the recommendations in this chapter focus on strengthening and accelerating efforts to achieve such breakthroughs. However, most of the recommendations do not assume that major breakthroughs are certain to occur. Indeed, a number of possible occurrences could set progress back: widespread drug resistance, social or political changes that increase vulnerability, or the refusal by decision-makers to allow evidence to guide programs and policies. For these reasons, planning for the future needs to have built-in flexibility and responsiveness to new knowledge as it emerges.

## 1. Build the knowledge base for long-term action

Increasing the durability of treatment programs and reducing the level of new infections to the point that the epidemic can eventually be eliminated requires developing new and better AIDS-fighting tools and strategies. Better knowledge also is required to focus prevention and treatment programs most effectively. Programs and policies should adapt based on new and emerging knowledge to stay relevant. Decision-makers at the national and subnational level require substantially better and more timely knowledge to respond effectively.

1. **Sustain financing for AIDS research and development**— Robust funding will continue to be needed to generate the tools and technologies required over the next generation. Key focus areas for research include development of a preventive vaccine, a cure, microbicides, pre-exposure prophylaxis, simpler diagnostic devices, and less costly treatment options.

2. **Adopt a new paradigm for vaccine research**—The existing paradigm for vaccine research, in which competing teams of research scientists work independently on vaccine candidates, often duplicating each other's work and failing to pursue other promising options, must be replaced. A new, collaborative, "open source" approach is needed, akin to the strategy successfully pursued with the Human Genome Project.

3. **Build the evidence base for longer-term programmatic and policy action to address key drivers of national and subnational epidemics**—Researchers should prioritize the development of user-friendly tools to assess, characterize, and understand key drivers of national and subnational epidemics. Increased investments in ethnographic and other social science research are needed to guide and evaluate the development of AIDS programs. All countries should undertake periodic assessments of incident HIV infections by modes of transmission and geographic distribution.

4. **Invest in both efficacy and effectiveness studies**—Prevention research must move from a single-minded focus on *efficacy* studies to include an emphasis on *effectiveness* in the real world. Greater investments are also needed to evaluate particular combinations of strategies.

5. **Implement prospective impact evaluations**—Evaluation should become a mandatory component of program design and implementation. Where possible, prospective impact evaluations should be built into programs, and results should be used to adapt programmatic and resource allocations. Impact evaluation efforts of programs funded by PEPFAR, the Global Fund, the World Bank, and other key funders should be aligned and coordinated. UNAIDS could help facilitate coordinated efforts in this area.

6. **Intensify development of incidence assays**—With the goal of strengthening impact evaluations, focused research is needed to develop affordable, reliable, user-friendly assays to measure HIV incidence.

7. **Invest in operational and translational research**—Focused research is needed to identify the factors that increase or decrease programmatic impact. Operational research is also needed to inform programmatic management of efficiency—for example, by generating optimal unit costs for well-run programs. Funders should invest in such studies.

8. **Establish research centers of excellence in developing countries**—Centers of excellence for research on AIDS and other priority health concerns should be established and locally led in low- and middle-income countries.

## 2. Give prevention the priority it deserves

Decision-makers must move from lip service to meaningful action on HIV prevention, prioritizing it as the mainstay of a sustainable response. The agreed goal for prevention policies and programs should be to maximize the number of infections prevented.

1. **Ensure strong leadership on HIV prevention**—If the number of new HIV infections in 2031 is to be sharply lower than it is

today, political leaders must build strong support for prevention programs and policies, even if they may show results only years down the road. Policies and practices that stigmatize and marginalize groups or individuals at high risk of infection must be avoided. Through visible leadership and adoption of sound programs and policies, political leaders should work to achieve concrete progress toward gender equality and the empowerment of women and girls.

2. **Alter the AIDS funding balance**—Financial support for HIV prevention must be increased to permit rough parity between prevention and treatment spending at the global level, taking into account country-level variations based on specific national needs. In implementing this recommendation, it is critical that donors and national decision-makers do not compromise treatment programs, as both treatment and prevention are essential and complementary components of an effective response.

3. **Implement combination prevention programs**—Every sound HIV-prevention plan reflects a strategic combination of behavioral strategies to promote risk reduction, biomedical interventions to reduce the likelihood that any episode of risk behavior will result in transmission, and social and structural approaches that minimize vulnerability and promote environments that are more conducive to HIV prevention. National plans need to describe how these different components fit together, justify the selection of interventions, explain how synergies will be captured and maximized, articulate concrete targets for results, and indicate projected causal pathways to achieve these results.

4. **Focus prevention efforts on the populations and settings where they are most needed**—Based on periodic assessments of the modes of transmission for new infections, decision-makers should select interventions and allocate resources to focus on the right mix of services in the right populations and geographic settings. Decision-makers should avoid the temptation to withdraw or not renew funding for services that are effectively helping to reduce rates of new transmission. Effective focusing of prevention services will simultaneously address emerging populations and sustain support for programs that are working.

5. **Implement an all-out prevention effort in Southern Africa**—In each hyperendemic country, the head of state should visibly lead a high-level prevention effort that maximizes the coverage and quality of HIV-prevention interventions. Major efforts should focus on the implementation and scaling-up of social change and community-level interventions. In hyperendemic countries, every adult citizen should know his or her HIV status and be appropriately supported in case of HIV infection.

6. **Ground prevention programs in the strengths of people living with HIV**—Prevention programs should undertake massive training and hiring of people living with HIV. Especially in high-prevalence settings, prevention efforts should invest in sustained campaigns to combine increased access to antiretroviral therapy, strong anti-stigma efforts, and a collective investment among all of society—including people living with HIV. To ensure local relevance, planning and priority-setting processes should involve civil society organizations and affected communities.

## 3. Ensure universal access to accessible, affordable, and sustainable treatment for people living with HIV

Historic achievements in expanding treatment access must not blind us to the reality that the current treatment model is not sustainable. Notwithstanding marked declines in drug prices, standard antiretroviral regimens remain too expensive and complex to make life-long therapy for tens of millions of individuals feasible in the most resource-limited settings. Additional strides are needed to improve treatment regimens and ensure their availability to all people living with HIV. In moving forward, the global community should adopt as its first priority extending life for the greatest number of people. To the greatest extent possible, consistent with maximization of treatment gains for people living with HIV, treatment strategies should maximize their contribution to reducing HIV transmission.

1. **More efficient regimens and treatment delivery strategies**—Urgent attention should focus on the development of regimens that are longer-lasting, simpler to take, less prone to resistance, and less costly than those currently being used. For example, single-pill

regimens or drugs that need to be taken only once a month rather than each day could significantly improve treatment adherence and delay the emergence of drug resistance. Efforts should be pursued to further lower the prices of antiretroviral regimens while maintaining economic incentives for continued biomedical research. All countries—and the global community as a whole—should pledge to achieve universal access to optimally effective first-line regimens for all people living with HIV. In addition, program managers and implementers, with the support of national governments, international donors, and technical agencies, should maximize efficiency in the delivery of treatment services.

2. **Diagnostics for resource-limited settings**—In addition to continuing efforts to expand access to existing diagnostic tools, intensive efforts should focus on the development of simple, affordable diagnostic tools. These include less expensive rapid tests for the diagnosis of HIV, as well as point-of-care tools that permit readings on key immunological and virologic indicators for the management of patients on antiretroviral therapy.

3. **Intensified focus on treatment adherence**—As an essential component of every treatment program, donors should amply fund intensive support for patient adherence, including but not limited to treatment buddies to help patients overcome impediments to adherence. Adherence interventions should build on robust evaluation research.

## 4. Implement a new code of conduct for the AIDS response

All AIDS stakeholders need to adopt new ways of working and new mechanisms of accountability to move from a short-term mindset to a longer-term perspective that requires local ownership, local capacity, and an evidence-informed and rights-based approach.

1. **Incentivize evidence-informed and rights-based programs and policies**—Strategies should specifically incentivize AIDS programs and policies to prioritize populations most at risk.

2. **Adopt a minimum legal framework**—All countries should adopt a minimum legal framework for the AIDS response (as discussed in Chapter 3, "Using knowledge for a better future"), consisting of legislation, such as

   • Decriminalize HIV status, transmission, and exposure
   • Decriminalize same-sex relationships/sexual practices
   • Guarantee equal rights of people living with HIV
   • Guarantee equal rights for men and women
   • Eliminate laws that limit access to health services for marginalized populations, including sex workers, people in same-sex relationships, and drug users
   • Decriminalize harm-reduction approaches for prevention of AIDS among those injecting drugs

3. **Support community-driven responses**—New mechanisms should be established for civil society review and approval of proposals to build community capacity and implement community-generated and driven AIDS programs. Particular attention should be given to community-based programs developed by and for key vulnerable populations, including but not limited to injecting drug users, men who have sex with men, sex workers, and young people. International donors should provide financial support for such programs.

4. **Build social capital to create AIDS-resilient communities**— Programmatic and policy responses should work to encourage a sense of agency, ownership, and responsibility about both individual and community responses to AIDS. Prevention and treatment models should be explicitly premised on the conceptualization of health as a public good and a human right, with a particular emphasis on addressing the needs of marginalized populations. Through focused funding, advocacy, and policy support, dynamic partnerships across organizations both within and outside the community should be facilitated, with the aim of building social capital. Mechanisms for local accountability that involve broad citizen participation should be supported, developed, and implemented.

5. **Build national capacity**—Funders should prioritize investments in educating new cadres of health professionals to manage AIDS and other health priorities. Rigorous monitoring indicators should be developed and implemented to assess achievements in building national capacity and guiding adaptation of strategies.

6. **Transition programs to local ownership**—As part of the new code of conduct, international nongovernmental organizations (NGOs) or other subcontractors that deliver AIDS programs should be mandated to include meaningful measures to build local capacity in their budgets and work plans, with the goal of eventually transferring primary responsibilities for all aspects of program implementation to local entities. Performance indicators to assess success in achieving these aims should be developed and implemented, and donors should make funding contingent on local stakeholders owning the programs.

7. **Develop local leadership**—The new code of conduct should require all programs to include specific strategies and budget lines to build and sustain local leadership to support the AIDS response.

## 5. Ensure robust, sustainable financing for a long-term response to AIDS

The world should transition from a funding paradigm that aims to cover AIDS-related costs to an investment paradigm that incentivizes quality, efficiency, and long-term planning and results.

1. **Renew and sustain global financial commitment**—Bilateral donors, national governments, philanthropic foundations, multilateral institutions, and other stakeholders should collectively commit to fund efficient, well-focused responses that have long-term horizons that aim for sharp declines in new infections and AIDS deaths over the next generation.

2. **Establish long-term budget horizons for long-term change**—Budget lines should be sufficiently robust to support substantial, long-term efforts and project cycles of 10–20 years.

3. **Prioritize low-income and high-prevalence countries**—The Global Fund, PEPFAR, and other leading donors should prioritize funding for the poorest countries, especially those with high HIV prevalence. Middle-income countries should assume the costs of their national AIDS programs.

4. **Improve program management to optimize effectiveness and efficiency**—Concerted efforts should focus on improving the management of AIDS programs. Incentives should be developed to encourage efficiency, including strategic integration of AIDS interventions with other services.

## 6. Exert leadership to achieve the aids2031 vision

The global attention span may sometimes be rather short, but the pandemic will remain a serious health threat for generations to come. Although its severity will vary tremendously from one country to another, it will continue to pose a historic test. Over the coming decades, AIDS will demand our attention, vigilance, and commitment.

1. **Institutionalize AIDS as a political issue**—In all high-prevalence countries, an annual parliamentary debate on AIDS should be mandatory, allowing national leaders to assess progress achieved, identify shortcomings, and plan for the future. In Southern Africa, AIDS should be a core cabinet matter, with active and well-monitored engagement of all relevant sector ministries. Planning horizons for national responses should be extended from 3–5 years to 10–20 years. At the global level, AIDS should remain a standing agenda item and a topic of ongoing debate and discussion at the United Nations and in the G20 forum. Leading regional political bodies should have permanent AIDS monitoring systems in place and convene at regular intervals to assess progress and address challenges.

2. **Structure reviews of national and subnational plans with a long-term view**—Yearly progress toward adopting long-term strategies, plans, and budgets should be reviewed.

3. **Keep the focus on AIDS**—The overall AIDS response must be much more closely linked with broader development efforts, yet it

must be recognized that the progress achieved thus far would never have occurred without the AIDS movement's singular focus on fighting the pandemic. AIDS shares some characteristics with other diseases and development priorities, but it is quite different in other ways. A new balance is needed that embeds the AIDS response in broader health and development efforts yet keeps a distinctive focus on addressing the unique and often politically and socially sensitive challenges of AIDS.

4. **Hold leaders accountable**—AIDS leaders have long decried punitive laws and policies that impede a sound response to AIDS epidemics and urged that funding be targeted to those who need services the most. Less often have the political leaders who adopt punitive policies or ignore most-at-risk populations been called to account. That must change. International bodies, civil society groups, the news media, and other stakeholders need to be more willing to criticize those who undermine effective action on AIDS. Leaders who fail to prioritize action for communities most at risk or who adopt punitive policies such as sodomy statutes or bans on harm-reduction services need to know that their actions have consequences.

5. **Broaden the AIDS coalition**—At global, regional, and national levels, AIDS stakeholders should work to broaden the coalition of AIDS supporters and champions to bring in a new generation of actors and to engage and work with advocates for other diseases, populations, and international development priorities.

6. **Strengthen watchdog functions**—Funders should significantly increase investment in independent civil society watchdog groups to monitor governments and other key stakeholders. As a general rule, watchdog groups should not include organizations that receive donor support for the provision of AIDS services.

Surprises and challenges will undoubtedly confront the world over the next generation. Yet despite these uncertainties, there are some clear actions that are needed today to change the face of aids by 2031. The choices in the coming years will determine the fate of millions of people.

# aids2031 working papers and additional resources

**(Available from http://www.aids2031.org/)**
**Brackets indicate working paper numbers**

## Leadership

*Working papers*

(1) Official government justifications and public ARV provision: A comparison of Brazil, Thailand, and South Africa

(2) A cross-country analysis of the determinants of antiretroviral drug coverage

(3) Networks of influence: A theoretical review and proposed approach to AIDS treatment activism

(4) Are country reputations for good and bad leadership on AIDS deserved?

(5) Transnational networks of influence in South African AIDS treatment activism

(14) Government leadership and ARV provision in developing countries

*Additional resources*

• Leadership spotlight: Gracia Violeta Ross (video)

## Science and technology

*Working papers*

(6)  aids2031 Science and Technology Working Group: A review of progress to date and current prospects

(7)  HIV eradication: Is it feasible?

(8)  Spurring innovation for the development of HIV/AIDS technologies

(9)  Are you experienced? Using the latest lessons learned from marketing research on consumer experience to improve the research and development of new HIV prevention technologies

(10)  Antiretroviral products for HIV prevention: Looking towards 2031

(11)  Community engagement in HIV prevention trials: Evolution of the field and opportunities for growth

(12)  Emerging trends in HIV pathogenesis and treatment

(13)  Increased access to diagnostic tests for HIV case management

*Final report*

Advancing Science and Technology to Change the Future of the AIDS Pandemic: A report from the aids2031 Science and Technology Working Group

## Financing

*Working papers*

(15)  The future role of the philanthropy sector fighting HIV & AIDS

(16)  Cost-effective interventions that focus on most-at-risk populations

(17)  Fiscal space and political space: Implications for aids2031

(18)  HIV and AIDS programs: How they support health systems strengthening

(19)  The evolution and future of donor assistance for HIV/AIDS

(20)  The past, present, and future of HIV/AIDS and resource allocation

(26) Assessing costing and prioritization in National AIDS Strategic Plans

(27) What works to prevent and treat AIDS: A review of cost-effectiveness literature with a long-term perspective

(28) The cost of antiretrovirals: Maximizing value for money

(29) Strategic planning in Honduras: A case study

(31) Estimating long-term global resource needs for AIDS through 2031

*Final report*

Costs and Choices: Financing the long-term fight against AIDS

## Social drivers

*Working papers*

(21) From risk-takers and victims to Young Leaders: Towards a different international AIDS response for young people

(22) Know your global crisis: What the AIDS industry might learn from the population story

(23) Re-thinking schooling in Africa: Education in an era of HIV & AIDS

(24) Addressing social drivers of HIV/AIDS: Some conceptual, methodological, and evidentiary considerations

*Final report*

Revolutionizing the AIDS Response: Building AIDS Resilient Communities

*Additional resources*

- Mobilizing social capital in a world with AIDS: Meeting report and recommendations
- The emergence of effective "AIDS Response Coalitions": A comparison of Uganda and South Africa
- Building AIDS competent communities: Possibilities and challenges

- Sex, rights, and the law in a world of AIDS—Meeting report and recommendations
- "This life is different": Street children's sexual realities & the APSA-Sexual Health Intervention Program
- Caught between the tiger and the crocodile: The campaign to suppress human trafficking and sexual exploitation in Cambodia
- Comprehensive LGBTQ-inclusive sexual health care for youth in state custody as a human right: The Teen SENSE Initiative
- Gender, HIV/AIDS, and the law in Zimbabwe
- Safe and consensual sex: Are women empowered enough to negotiate?
- Men's gender inequality perceptions influence their higher-risk sex in northern India
- Jessica Ogden, Understanding and Addressing Structural Factors in HIV Prevention, 2008 Mexico AIDS Conference (Video)

## Programmatic response

*Working papers*

(25) Is AIDS exceptional?

*Final report*

Making choices, embracing complexity, driving and managing change: The HIV programmatic response over the next generation

*Additional resources*

- November 2008 Constituency Consultation—Meeting recommendations

## Communication

(30) Future connect: A review of social networking today, tomorrow, and beyond and challenges for AIDS communicators

## Hyperendemic areas

*Final report*

Turning off the tap: Understanding and overcoming the HIV epidemic in Southern Africa

## Countries in rapid economic transition

*Final report*

Asian economies in rapid transition: HIV now and through 2031

## Modelling

*Additional resources*

- Geoff Garnett, interview at the Mexico AIDS Conference (video)

# About the authors

**The aids2031 Consortium** includes nine thematic working groups on the topics of modeling, science and technology, social drivers, the programmatic response, financing, communication, leadership, a special look at hyperendemic countries (Southern Africa), and countries in rapid economic transition (focusing on China, India, Indonesia, Malaysia, and Thailand). These nine groups, along with a group of aids2031 young leaders, engaged over 500 people around the world in discussions, debates, and issue-specific analyses on the current and future state of AIDS.

The work of the aids2031 Consortium and its working groups is led by the **Steering Committee** listed here:

> **Zackie Achmat**, Founder, Treatment Action Campaign, South Africa
>
> **Ricardo Baruch**, Global Youth Coalition on HIV/AIDS Taskforce, Mexico
>
> **Stefano Bertozzi**, Director, HIV and Tuberculosis, Global Health Program, Bill & Melinda Gates Foundation, and Chair of the aids2031 Steering Committee
>
> **Myung-Hwan Cho**, President, AIDS Society of Asia and the Pacific, and Professor, Konkuk University, South Korea
>
> **Achmat Dangor**, CEO, Nelson Mandela Foundation, South Africa
>
> **Paul Delay**, Deputy Executive Director, UNAIDS, Switzerland
>
> **Alex deWaal,** Program Director, HIV/AIDS and Social Transformation, Social Science Research Council, USA
>
> **Chris Elias**, President, Program for Appropriate Technology in Health (PATH), USA
>
> **David de Ferranti**, Executive Director, Global Health Initiative, Brookings Institution, USA

**William Fisher**, Director, Department of International Development, Community and Environment, Clark University, USA

**Geoffrey Garnett,** Professor, Imperial College London, UK

**Denise Gray-Felder**, President, Communication for Social Change Consortium, USA

**Geeta Rao Gupta**, Senior Fellow at the Bill & Melinda Gates Foundation and former President, International Center for Research on Women, USA

**Rob Hecht**, Principal and Managing Director, Results for Development, USA

**Heidi Larson**, Executive Director aids2031 and Senior Lecturer at the London School of Hygiene and Tropical Medicine, UK

**Callisto Madavo**, Professor, Georgetown University, USA

**Michael Merson**, Director, Duke University Global Health Institute, USA

**Sigrun Mogedal**, Ambassador on HIV/AIDS, Ministry of Foreign Affairs, Norway

**Prasada Rao**, Senior Advisor to the Executive Director of UNAIDS and former UNAIDS Regional Director for Asia and the Pacific

**Leonardo Simao**, Chief Executive, Joachim Chissano Foundation, Mozambique

**As Sy**, UNICEF Regional Director for East and Southern Africa, and former UNAIDS Director of Partnerships and External Relations

**Ex officio members:**

**Robert Hemmer**, National Service of Infectious Diseases, Centre Hospitalier de Luxembourg

**Peter Piot**, Director of the London School of Hygiene and Tropical Medicine, and former UNAIDS Executive Director

**Todd Summers**, Senior Advisor for Global Health at ONE, formerly Senior Policy Officer for Global Health, Bill & Melinda Gates Foundation, USA

**Writing team:**

Stefano Bertozzi, William Fisher, Michael T. Isbell, Lindsay Knight, Heidi Larson, and Peter Piot

# Acknowledgments

Eduard Beck, Paul Bekkers, Kelsey Case, Caitlin Chandler, William Dowell, Jay Dowle, Carly Edwards, Olavi Elo, Eileen Garred, Edward Girardet, Robert Goble, Timothy Hallett, Anne Hendrixson, Kate Nightingale, Owen Ryan, Denise Searle, Karen Stanecki, Jeff Sturchio, Barbara Thomas-Slayter, Alan Whiteside, and Marijke Wijnroks.

Additional thanks to the institutional funders and supporters of the aids2031 Consortium—AmfAR (American Foundation for AIDS Research), Bill & Melinda Gates Foundation, Ford Foundation, Galileo Global Advisors, Google.org, Grand Duchy of Luxembourg, Irish Aid, Merck & Co. Inc., the Packard Foundation, Tiller LLC, UNAIDS, UNDP, UN Foundation.

# Index

**FT** Press

FINANCIAL TIMES

SCIENCE

The life sciences revolution is transforming
our world as profoundly as the industrial
and information revolutions did
in the last two centuries.
FT Press Science will capture the excitement and
promise of the new life sciences, bringing breakthrough
knowledge to every professional and interested citizen.
We will publish tomorrow's indispensable work in
genetics, evolution, neuroscience, medicine,
biotech, environmental science, and whatever
new fields emerge next.
We hope to help you make sense of the future,
so you can *live* it, *profit* from it, and *lead* it.